Restoring Alaska:
Legacy of an Oil Spill

ALASKA GEOGRAPHIC®
Volume 26 Number 1 / 1999

To teach many more to better know and more wisely use our natural resources

EDITOR
Penny Rennick

ASSOCIATE EDITOR
Rosanne Pagano

PRODUCTION DIRECTOR
Kathy Doogan

MARKETING DIRECTOR
Jill S. Brubaker

BOOKKEEPER/DATABASE MANAGER
Claire Torgerson

OFFICE ASSISTANT
Melanie Britton

POSTMASTER:
Send address changes to:

ALASKA GEOGRAPHIC®
P.O. Box 93370
Anchorage, Alaska 99509-3370

PRINTED IN U.S.A.

ISBN: 1-56661-046-X

PRICE TO NON-MEMBERS THIS ISSUE: $21.95

COVER BACKGROUND: *Oil shimmers on a beach in Northwest Bay in western Prince William Sound.* (Patrick J. Endres)
COVER INSET, LEFT: *A sea otter passes time in Valdez.* (Alissa Crandall)
COVER INSET, CENTER: *Three-year-old Mikel Grillo shows off a razor clam scooped from a beach at Clam Gulch on Cook Inlet.* (Al Grillo)
COVER INSET, RIGHT, AND PREVIOUS PAGE: *A pelagic cormorant finds temporary shelter from spilled oil in 1989 at a bird rescue center in Valdez.* (Alissa Crandall)

FACING PAGE: *Government studies have detected oil spill damage to seaweed. On the accident's 10th anniversary, Gov. Tony Knowles said, "The legacy of this spill will be about people working together, to restore the injured environment and to prevent anything like this from every happening again."* (Natalie B. Fobes)

ALASKA GEOGRAPHIC® (ISSN 0361-1353) is published quarterly by The Alaska Geographic Society, 639 West International Airport Rd., Unit 38, Anchorage, AK 99518. Periodicals postage paid at Anchorage, Alaska, and additional mailing offices. Copyright © 1999 by The Alaska Geographic Society. All rights reserved. Registered trademark: Alaska Geographic, ISSN 0361-1353; key title Alaska Geographic. This issue published April 1999.

THE ALASKA GEOGRAPHIC SOCIETY is a non-profit, educational organization dedicated to improving geographic understanding of Alaska and the North, putting geography back in the classroom and exploring new methods of teaching and learning.

MEMBERS RECEIVE ALASKA GEOGRAPHIC®, a high-quality, colorful quarterly that devotes each issue to monographic, in-depth coverage of a specific northern region or resource-oriented subject. Back issues are also available. Membership is $49 ($59 to non-U.S. addresses) per year. To order or to request a free catalog of back issues, contact: Alaska Geographic Society, P.O. Box 93370, Anchorage, AK 99509-3370; phone (907) 562-0164, fax (907) 562-0479, e-mail: akgeo@aol.com.

SUBMITTING PHOTOGRAPHS: Those interested in submitting photos for possible publication should write for a list of upcoming topics or other photo needs and a copy of our editorial guidelines. We cannot be responsible for unsolicited submissions. Submissions not accompanied by sufficient postage for return by certified mail will be returned by regular mail.

CHANGE OF ADDRESS: The post office will not automatically forward *ALASKA GEOGRAPHIC®* when you move. To ensure continuous service, please notify us at least six weeks before moving. Send your new address and membership number or a mailing label from a recent issue of *ALASKA GEOGRAPHIC®* to: Alaska Geographic Society, Box 93370, Anchorage, AK 99509. If your book is returned to us by the post office because it is for some reason undeliverable, we will contact you to ask if you wish to receive a replacement for a small fee to cover additional postage.

COLOR SEPARATIONS: Graphic Chromatics

PRINTING: Hart Press

The Library of Congress has cataloged this serial publication as follows:

Alaska Geographic. v.1-
    [Anchorage, Alaska Geographic Society] 1972-
    v. ill. (part col.). 23 x 31 cm.
    Quarterly
    Official publication of The Alaska Geographic Society.
    Key title: Alaska geographic, ISSN 0361-1353.

    1. Alaska—Description and travel—1959-
    —Periodicals. I. Alaska Geographic Society.

F901.A266   917.98'04'505   72-92087

Library of Congress   75[79112]   MARC-S.

ABOUT THIS ISSUE:
Restoring Alaska: Legacy of an Oil Spill tells the decade-long story of the Exxon Valdez oil spill, focusing on spill-zone lands and scientific research as well as the people whose lives were touched by the tanker accident in 1989. We are indebted to the scientists, industry and government officials and oil spill researchers coast to coast who contributed to this book and shared our goal of explaining oil spills and oil spill science to a wider audience.
ALASKA GEOGRAPHIC® expresses special thanks to Exxon senior media officer Edward Burwell; Lydia Chase and James McFarlane of the *Seattle Times*; Joe Hunt, media coordinator, Exxon Valdez Oil Spill Trustees; Carrie Holba and staff of the Alaska Resources Library and Information Services; and Stan Jones, public information manager, Prince William Sound Regional Citizens' Advisory Council.

Contents

Foreword: Legacy of an Oil Spill 5
   By Sherry Simpson

Restoring Alaska: Science Follows Oil's Wake ... 11
   By Kris Capps

■ **Injured Species and Resources at a Glance** 30
■ **Following the Flow in Prince William Sound** 39

Healing Lands: Alaska Invests in Nature 45
   By Natalie Phillips

■ **Map-making Scientists Guide Oil Spill Cleanup** ... 50
   By Greg Chaney
■ **View From a Tide Pool** 60
   By Charles P. Wohlforth

Out of the Sound: People of the Oil Spill 63
   By Rosanne Pagano

■ **Clipped Hair, Spilled Oil** 66
■ **Home at Last: Museum Captures Kodiak's Past** .. 74

Lessons of an Oil Spill: Alaska and Beyond 83
   By Ernest Piper

■ **Then and Now: Progress in the Sound** 85

When Oil Hits Water: A Spill Chronology 102

Bibliography 109

Index 110

Legacy of an Oil Spill: 'Move on, Yes; Forget, Never'

By Sherry Simpson

 In the decade since the Exxon Valdez collided with history, the nation's largest oil spill has been distilled into a series of calculations: Crude oil spilled (11.2 million gallons). Coastline oiled (1,300 miles). Sea otters killed (2,800). Birds dead (250,000). Cleanup workers employed (11,000). Oil recovered (about 12 percent). Civil settlement paid by Exxon ($900 million). Criminal penalty exacted by jury ($5 billion, under appeal). Species of birds, mammals and fish significantly damaged (26). Species "recovered" (2).

So much is beyond measure, though. Some equations will not resolve. No one has yet fixed the true value of an otter, or a fouled beach, or a lifestyle lost. How to reckon the anger, the sorrow, the regret?

FACING PAGE: Nearly a decade of spill zone research compiled by government and industry scientists is "a legacy of hope," one agency scientist says, because researchers now know how to monitor environnmental damage. Knight Island's southwest flank is glimpsed across Long Channel from Squire Island, on the Sound's west side. (Chlaus Lotscher)

Humanity is not included on the list of species damaged by the oil spill and so the question remains: Prince William Sound may indeed recover some day, but will we?

Ten years after the oil spill, ask anybody, "What is the legacy of the Exxon Valdez?" Chances are you'll hear, as I did, either puzzled silence or knee-jerk answers:

"A lot of people made money."

"Greed and avarice."

"Isn't it all fixed now?"

I'd like to believe that people answer this way because what they mean to say is "I don't know." So much remains in flux. Courts grind away, decisions are appealed or reversed. Scientists wrangle with difficult questions and sometimes they wrangle with each other. Ecosystems are changing in the Gulf of Alaska, but which changes were caused by the oil spill and which by shifts in climate and food sources? Faced with such complications, how easy it becomes to succumb to complacency again.

Similarly, the passage of time has rendered the events of the oil spill more perplexing than ever. Once it seemed simple to identify the heroes, the villains, the victims. But the smudge of complicity lingers, shared by a ship's crew that faltered at the helm, the companies and agencies that did too little and much too late, the watchdogs who failed to watch, the citizens distracted by easy money, the consumers who demanded more, more, more.

Some stains cannot be removed entirely no matter how hard we rub. The spill created every kind of damage imaginable: creatures died by the hundreds of thousands, fisheries were impaired, habitats corroded, trust lost. The lives of many people — villagers, fishermen, even the cleanup workers — were grounded with the Exxon Valdez too. A decade ago, wholesale ruination seemed inevitable.

Yet hasty judgments have been confounded. The spill's consequences seem neither as lasting as once feared, nor as fleeting as some would have us believe. Look out upon the waters of Prince William Sound and all seems serene, a landscape restored. The persistence of life appears miraculous. But dig your hands below the surface of some beaches and oil will blacken them, a constant reminder that the vitality of Prince William Sound has dimmed in ways we cannot clearly see.

It would be wrong, though, to ignore the good things that have emerged over the

FACING PAGE: *Awaiting settlement of their claims 10 years after the Exxon Valdez spill are thousands of Alaskans, including some for whom commercial fishing is a family business. In Whittier, a Prince William Sound community untouched by oil, a couple mends nets for a bowpicking boat. (Patrick J. Endres)*

ABOVE: *A woman weeps while flying over the slick in a small plane in 1989. "A decade ago, wholesale ruination seemed inevitable," essayist Sherry Simpson observes, noting too that progress is part of Alaska's oil spill lesson. (Roy Corral)*

ABOVE RIGHT: *Surfbirds use the Prince William Sound region as a rest stop and feeding area during spring migration. Wildlife managers quickly sized up the spill's potential for harm to migrating flocks. (Patrick J. Endres)*

years. Nearly 760,000 acres of land, much of which would have been logged otherwise, have been purchased or protected by the Exxon Valdez Oil Spill Trustee Council with settlement money. The new Alaska SeaLife Center in Seward allows visitors to glimpse both the oceanic and scientific worlds in the only major cold water marine lab in the world. Research continues to expand knowledge about coastal Alaska's ecology and the behavior of oil once it's been loosed upon the world. Federal and state legislation created tougher tanker regulations, stiffer penalties and a new focus on disaster prevention and response.

The spill taught us about ourselves, too. Nobody can be sure how he or she will behave in a disaster. Now we know. Some

of us turn to each other (remember the fishermen's mosquito fleet to save a hatchery in the "Battle of Sawmill Bay"); some of us turn on each other (think of the carping over who has reaped money from the spill and who has not).

Now we seize eagerly upon the faintest signs of healing. Bald eagles and river otters appear to have regained their original strength. Several other populations seem to be mending, including common murres, sockeye and pink salmon, and denizens of the tidal zones. Native villagers have returned cautiously to eating traditional foods offered by the ocean.

A real danger lies in the false sense that recovery is certain, though. There will never come a day when some potentate announces,

"The oil spill is over! It's all better now!" Nor will we ever be free from jeopardy.

As long as we require petroleum, oil spills will afflict us. Within 30 months after North Slope crude gushed into Alaska's waters, four spills of more than 10 million gallons occurred throughout the world. As I write this, a tanker grounded off the Oregon coast has been set afire to prevent more oil from washing ashore.

A spring sun rises on Prince William Sound. (Patrick J. Endres)

Perhaps the true legacy is our sense of vulnerability, the bitter awareness of all that can go wrong and all that can be forfeited. We saw how easily the richness of life can slip away, but the magnitude of the oil spill's devastation dominates our vision so that we do not always see the losses that occur around us daily, the steady erosions of ecosystems and landscapes and a sense of place. Disasters do not always arrive in large-scale formats or in made-for-TV movies. They can happen drop by drop, forest by forest, bay by bay, otter by otter.

Today some people — usually people who did not witness those dreadful days — say to Alaskans, "Get over it. It's been 10 years." But the Exxon Valdez oil spill is nothing people can get over that easily — nor should they. Move on, yes. Make changes, certainly. Forget, never. ✈

Journalism instructor and ALASKA GEOGRAPHIC® *contributor Sherry Simpson teaches at the University of Alaska Fairbanks. Her book,* The Way Winter Comes, *is a collection of essays about Alaska published in 1998.*

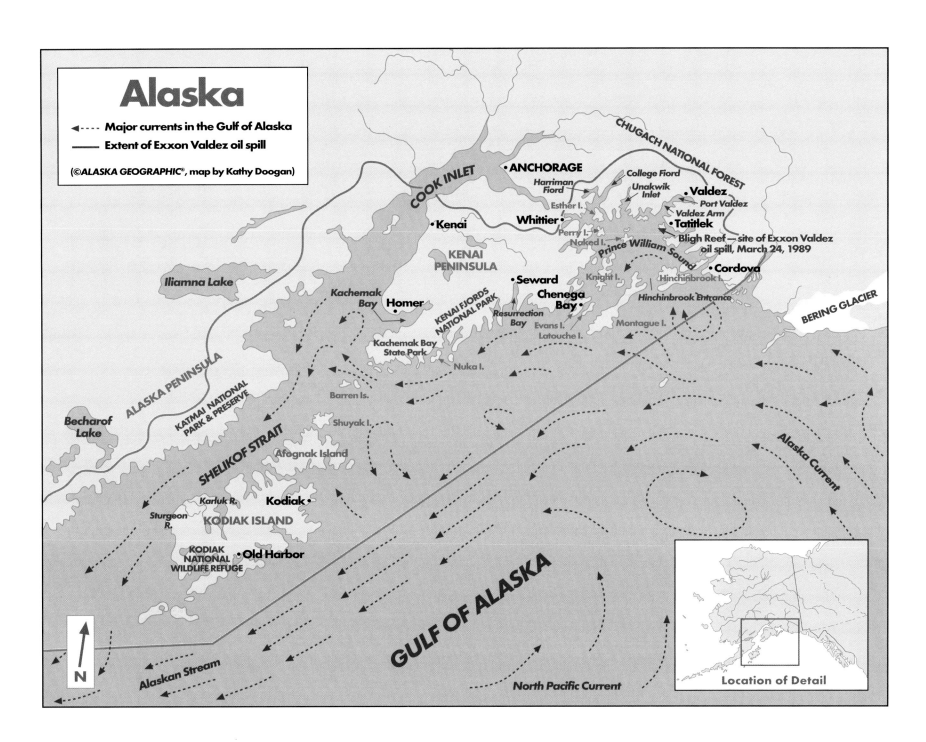

Alaska

◄ - - - - Major currents in the Gulf of Alaska

──────── Extent of Exxon Valdez oil spill

(©*ALASKA GEOGRAPHIC*®, map by Kathy Doogan)

CHUGACH NATIONAL FOREST

COOK INLET

•**ANCHORAGE**

College Fiord

Harriman Fiord

Unakwik Inlet

•**Valdez**

Port Valdez

Valdez Arm

Esther I.

•**Kenai**

Whittier •

•**Tatitlek**

Perry I.

Naked I.

Bligh Reef — site of Exxon Valdez oil spill, March 24, 1989

Prince William Sound

KENAI PENINSULA

Iliamna Lake

Knight I.

•**Cordova**

Hinchinbrook I.

Kachemak Bay

Homer •

•**Seward**

Chenega Bay •

Resurrection Bay

Hinchinbrook Entrance

Montague I.

BERING GLACIER

ALASKA PENINSULA

KENAI FJORDS NATIONAL PARK

Evans I.

Latouche I.

Kachemak Bay State Park

Nuka I.

KATMAI NATIONAL PARK & PRESERVE

Becharof Lake

Barren Is.

Shuyak I.

Afognak Island

Alaska Current

SHELIKOF STRAIT

Karluk R.

Kodiak •

Sturgeon R.

KODIAK ISLAND

KODIAK NATIONAL WILDLIFE REFUGE

•**Old Harbor**

GULF OF ALASKA

N

Alaskan Stream

- - - - - **North Pacific Current** - - - -

Location of Detail

Restoring Alaska: Science Follows Oil's Wake

By Kris Capps

Scientist Daniel Esler cradled a dusky brown harlequin duck in his arms as the female, wrapped in a cotton towel and weighing little more than one pound, slowly emerged from an anesthetic fog. The bird's head rested quietly against Esler's torso. For this little harlie, as Esler's team dubbed the sea duck, it had been quite a day.

After flying into a mist net strategically set by Esler and his research crew, the harlequin was gently extracted from the tangle and carried aboard a research vessel. Scientists there weighed and measured the bird, noted its age and sex and collected a blood sample. They placed a numbered band around one leg.

"Sure is a scruffy looking bird," said Esler, the project leader, who handed the captive bird to wildlife veterinarian Dan Mulcahy. Speaking up for the duck, technician Ellie

Mather bantered back: "Sure is a scruffy-looking bander."

Drugged by an anesthetic hood over its head, the bird underwent surgery. Mulcahy cut a tiny incision in its abdomen, removed a teardrop-sized slice of liver and sutured the cut. He handed the harlequin back to Esler, who made sure it recovered before release to the wilds of Prince William Sound.

More than 200 harlequin duck carcasses were found in 1989 following the Exxon Valdez oil spill. Most losses were in Prince William Sound, a wildlife cradle offering forage, nesting and shelter to species up and down the food chain. Experts estimate that carcasses retrieved represent only a portion of harlequins lost; the spill zone population has yet to rebound, government scientists say.

Esler's research in the spring of 1998 is part of ongoing studies to track the health of Prince William Sound harlequins. Liver samples, for instance, are examined for the presence of an enzyme produced by harlequins exposed to contaminants. Esler's study shows that harlequin ducks in the spill zone had a much higher concentration of the telltale enzyme than harlequins from oil-free areas.

By monitoring harlequin ducks in both oiled and oil-free regions, Esler, who works for the Biological Resources Division of the U.S. Geological Survey, also studies whether spilled oil still is affecting the birds. His finding: Survival rates for adult females were lower in oiled areas. The good news for harlequins, Esler said, is that ducks in both areas seemed to have plenty of food, similar body mass and similar blood chemistry. But he predicts recovery will take a long time.

Harlequins "invest relatively little" in reproduction each year, Esler notes. "Unlike a mallard, they can't dramatically increase numbers over a year or two." Harlequins, which feed in intertidal and subtidal zones where oil strands formed in 1989, also appear to be faithful to certain sites, even those marred by oil.

Oil from the Exxon Valdez tanker gushed into pristine waters that were unusually calm for three days until strong winds churned the slick into "mousse," a sea water and oil emulsion that worked its way southwest.

Scientists estimate that 35 percent of spilled oil evaporated and 25 percent reached the Gulf of Alaska, where it was driven out to sea or beached; the remaining

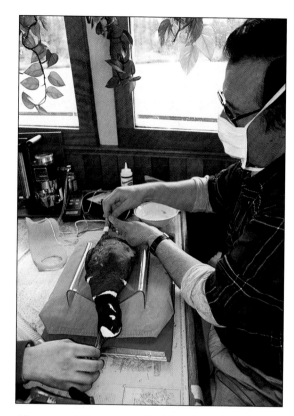

40 percent deposited itself on beaches in Prince William Sound.

Observers in 1989 calculated that more than 700 miles of shoreline were oiled, including 200 miles of shore classified as "heavily oiled." Beyond the Sound, more than 2,400 miles of shoreline in the Kenai Peninsula-Kodiak region also were oiled. For harlequin ducks like the one resting in Esler's arms, pollution came ashore where birds live and feed. More than 36,000 oiled bird carcasses of various species were retrieved in the spill area while estimates of total bird losses range to 250,000 individuals, largest of any damaged species. The spill claimed 2,800 sea otters, 300

harbor seals, 250 bald eagles and untold mussel beds, clams and miniature marine life smothered by crude.

Scientists, outdoor enthusiasts, fishermen and Alaska Natives who harvest Prince William Sound for traditional food all worried that Alaska waters might absorb spill "aftershocks" for years if oil seeped into beaches, contaminated food sources and disrupted marine life cycles. Molly McCammon, executive director of the Exxon Valdez Oil Spill Trustee Council, which oversees Alaska's settlement fund, says the spill zone's ecosystem is "well on it way" to recovery. But, she notes, "long-term impacts on individual populations may take decades to fully heal."

Among species yet to recover, government scientists say, is a closely studied group of Prince William Sound killer whales that lost 13 of its members. Eleven whales were lost from another pod. Prince William Sound's herring fishery collapsed following the spill; scientists remain unclear whether Exxon Valdez oil played a part. The Sound's pink salmon fishery also went into decline, and

At far left, veterinarian Dan Mulcahy collects a tissue sample from a harlequin duck in 1998. Harlequins are named for the male's distinctive markings. Left, researcher Daniel Esler releases one of the birds after data is collected to help determine the harlequin's post-spill numbers. The bird is among species deemed not recovered from spill losses, trustees say. (Both: Patrick J. Endres)

scientists report increased egg mortality in oiled streams. Current research is aimed at determining whether pink salmon suffered genetic damage and are passing a contaminated "inheritance" to future generations.

After a decade of science, government researchers believe a new legacy of the spill has emerged — oil as a slow-acting poison whose toxic components remain in the environment longer. Scientists now say that the amount of oil in water capable of causing genetic damage in pink salmon is just 1 part per billion, one order of magnitude lower than the 15 parts per billion in water permitted under state standards for polycyclic aromatic hydrocarbons, the most toxic class of compounds in oil.

"The toxicity is more persistent than we ever imagined," said Juneau-based biologist Ron Heintz of the National Marine Fisheries Service, who studied oil's long-term effects on pink salmon.

Exxon-funded research tells a different story. The company, which also has spent the past 10 years collecting scientific data in the spill zone, released findings in March 1999 stating that pink salmon, sea otters and several sea bird populations appeared more resilient to spilled oil than originally thought. Early suggestions of "negative impacts may have been unfounded," University of Wisconsin biologist John A. Wiens said, adding that pink salmon and many sea birds had exhibited no obvious detrimental impacts; sea otters and some birds had sustained spill impacts "followed by clear evidence of subsequent recovery."

"The apparent resilience of these species (may) be related to the high natural variability of these ecosystems," Wiens concluded.

To Know a Spill Zone: Research Looks Ahead and Back

Of the 26 species and resources deemed by the trustees to have been injured by the oil spill, only two — bald eagles and river otters — now are classified "recovered."

BELOW: *Although not on injured species lists, a white winged scoter is tagged for post-spill population studies in Prince William Sound. (Patrick J. Endres)*

RIGHT: *Cleansed of oil, a sea otter awaits release in 1989. (Al Grillo)*

Bald eagles made the list in 1996, river otters were added in 1999.

On the "recovering" list are Pacific herring, pink salmon, sea otters, black oystercatchers, marbled murrelets and mussels among others; "not recovering" are harlequin ducks, harbor seals, killer whales and common loons. Still other wildlife fall into the "recovery unknown" category, including Dolly Varden, cutthroat trout, Kittlitz's murrelet and rockfish. Creatures on the unknown list were little studied in the past and current research either is inconclusive or incomplete.

After crude oil entered the Sound, the entire marine system became a unique

laboratory for government and Exxon-funded scientists alike. Settlement funds underwrote new government research focusing on birds, mammals and fish as well as archaeology and the habits of the ocean itself. New studies on wind and ocean currents, for example, can help predict the path of a future oil spill and the intensity of plankton bloom, a foundation of the marine food chain.

In Prince William Sound, that chain includes Pacific herring and pink salmon, two species at risk after the spill and traditionally top money fish for the commercial fleet. Because abundant data on local pink salmon and herring stock was

Roughly one-fifth of the 840 cormorant carcasses retrieved after the spill were identified as red-faced cormorants. Based on surveys through 1998, the birds are listed by oil spill trustees as not recovered from spill losses. (Patrick J. Endres)

at hand, some atypical observations in the spill zone could be linked to the presence of oil. For instance, based on pre-spill knowledge, researchers knew that reduced growth of salmon in oiled estuaries indicated they were unlikely to survive as well as counterparts from clean waterways.

But there was much more that wasn't known about spill zone waters and wildlife, putting researchers at a disadvantage as they tried to devise before-and-after comparisons. Research was further complicated because Prince William Sound is a varied ecosystem, so that the health and population size of some species depend on when and where in the Sound they reside.

Some spill science remains in dispute a decade after the accident. Exxon Corp., which has appealed a $5 billion verdict following an oil spill class action heard in Anchorage, says it believes the Sound has recovered. Agencies, environmentalists and some Native villages dispute that, citing residual oil lingering at certain sites, mostly in western Prince William Sound and the Kenai Peninsula's southern tip. A settlement that resolved government lawsuits called on Exxon to pay $900 million over a 10-year period to restore the environment. That fund is administered by the oil spill trustees, a six-person panel representing government and state agencies.

The federal criminal plea agreement

required Exxon to pay a fine of $250 million. Money went into two restitution funds of $50 million each, administered separately by federal and state governments; $125 million was forgiven in recognition of Exxon's cooperation during the cleanup, its timely payment of many private claims and environmental precautions taken during the spill; $12 million went to the North American Wetlands Conservation Fund and $13 million went to a Victims of Crime Act account.

Major projects funded by the civil settlement and overseen by the oil spill trustees include:

• **The Alaska Predator Ecosystem Experiment**, known as APEX. This project tests the general assumption that low food abundance contributes to the decline of sea bird and marine mammal populations in Prince William Sound and Cook Inlet.

• **The Nearshore Vertebrate Predator Project**. Involving a number of research projects and several different species, the project assesses whether populations are recovering, factors limiting recovery, and how to promote recovery. Species under study include sea otters, river otters, harlequin ducks and pigeon guillemots.

• **The Sound Ecosystem Assessment Project**, examining how ecosystem health influences early lives of pink salmon and Pacific herring. Juvenile fish are studied to see how survival is linked to food availability or predators. Research includes how food is carried by ocean currents, ocean

Spilled oil leaves a "bathtub" ring on some Prince William Sound shores in 1989. (Roy Corral)

LEFT: *In its 1999 assessment, Exxon noted that Prince William Sound and the wider spill zone had probably seen more environmental study than anywhere in the world. Sea stars, familiar to tidal pool gazers, were not overlooked. (Al Grillo)*

ABOVE: *Among resources classified in 1999 as recovering are the Sound's intertidal and subtidal zones as well as sediments under cobble and boulder beaches, found throughout the spill zone. (Patrick. J. Endres)*

climate and modeling to study ecosystem stress.

• **Habitat investigations pursued by Auke Bay Fisheries Laboratory**, a unit within the National Oceanic and Atmospheric Administration. Among other things, scientists study any long-term effects of remaining oil pockets and research long-term effects of oil on reproduction of herring and pink salmon. The Juneau-based lab conducts monitoring and restoration research for the trustee council.

The spill also spawned the Prince William Sound Science Center in Cordova, a nonprofit research and education group founded in 1989 for long-term ecosystem studies in Prince William Sound.

Located in the cozy fishing village of Cordova, the center presents science to local residents by hosting international workshops on regional issues and public lectures where residents are encouraged to meet researchers. The center brings science to classrooms throughout the Sound and administers the Oil Spill Recovery Institute, an agency set up by Congress in 1990 to find ways to prevent and respond to oil spills in cold waters.

Research done in the spill zone laboratory may help improve cleanup of future oil spills or prevent them completely. Studies are producing a clearer understanding of the marine ecosystem, research that will help fisheries managers and other policy-makers.

"Ten years ago, no one could answer questions about what the long-term effects of this spill would be," says McCammon, the trustees council executive director. "Today, we can look back and see that in western Prince William Sound, oil remains on some beaches, sea otters are not re-populating once-oiled areas, Pacific herring suffered a complete collapse and are just now rebuilding to harvestable levels, and the people who live, work, and play in the region continue to live with the impacts of spilled oil."

Tenacious Oil, Stubborn Questions

Most visitors captivated by Prince William Sound will never notice lingering oil. In some spots it can be smelled before it's seen, but that is rare. In most places, seekers of residual oil have to know where to look — under boulders on a particular

BELOW: *How clean is clean? One year after cleanup was declared complete in 1992, a visit to western Prince William Sound's Eleanor Island turns up subsurface oil. Residual oil persists in isolated spots today. (Patrick J. Endres)*

BELOW RIGHT: *Government research into the Sound's pigeon guillemot population shows the birds, known to feed in nearshore waters, have yet to recover from the oil spill. Like the marbled murrelet, pigeon guillemot numbers appear to have been declining before the spill for reasons that aren't yet clear. (Kevin Hartwell)*

beach, deep in gravel or under beds of mussels. Exxon, which presented its 10-years-after findings at a March 1999 conference in Seattle, has said the Sound is "essentially" recovered, noting that researchers disagree about the Sound's pre-spill health as well as when to declare a species recovered.

But scientist Ron Heintz with the National Marine Fisheries Service says oil leftovers are land mines, hidden from view yet primed to do harm with a slow drip of contamination. "Once the oil is exposed to the surface, it can leach toxic chemicals into the environment as it undergoes the natural cleaning process," Heintz says.

Some authorities believe any ongoing contamination will continue to mar Prince

William Sound, from mussels along the shores to killer whales that pass contamination to offspring in mother's milk. "Whether it (residual oil) is having a biological effect will be debated endlessly," said wildlife biologist Dan Rosenberg, who studies harlequin and scoter ducks. Another factor in that debate: How, or if, Alaska's warmer climate — as much as 5 degrees have been gained over the past 30 years — is altering ecosystems such as Prince William Sound. Ongoing research into the region's changing food web could provide

Exxon on Recovery

Here is Exxon's statement on species recovery in Prince William Sound:

"We believe the definition of recovery being used by the trustees is misleading because it is not a practical or accurate measure. Recovery of the Prince William Sound ecosystem cannot be measured and defined by the recovery of the few species the trustees are investigating. (The ecosystem) is populated by thousands of other species that were not impacted by the spill, or were impacted but recovered quickly.

"Recovery means a healthy biological community has been reestablished and that the plants and animals characteristic of that community are present and are functioning normally. This definition works because (it) is not dependent on pre-spill vs. post-spill measurements; it recognizes that nature changes all the time, a concept that scientists call 'natural variability.'

"Any definition that uses as an absolute standard the return to pre-spill conditions just isn't realistic." •

Memories of Prince William Sound: At left, adventure seekers soak up the Sound's beauty by sailboat while package tours funnel thousands of tourists each season through Prince William Sound communities of Valdez, Cordova and Whittier. Buses, facing page, are hauled by the Alaska Railroad to Whittier, a jumping-off spot for tour boats exploring the Sound. (Both: Alissa Crandall)

answers. "Scientists talk about the Exxon Valdez oil spill as a 'pulse perturbation' — a disturbance that was short-lived, albeit catastrophic," says University of British Columbia researcher Thomas Okey.

As recently as 1997, residents of the Alaska Native village of Chenega Bay found themselves confronting tenacious oil. At the villagers' urging, another round of cleanup was scheduled at Sleepy Bay, on the heavily oiled north end of Latouche Island, and on the northeast corner of Evans Island. Chenega villagers said that despite the passing years, they still hesitated to hunt or fish because of remaining oil. A 25-member crew made up primarily of Chenega residents worked 36 days in the summer of 1997, trying to remove crude oil from 2.5 acres of beach.

They loosened gravel with compressed air and a cleaning agent, then flooded the worksite with seawater. Oily water washed down the beach where floating booms trapped it and absorbent pads sopped it up. Monitoring of the beach has shown that while a lot of oil was removed in 1997, some still remains buried deep under boulders, hidden from wind, weather and natural cleansing.

Seeking Biological Cause and Effect:
A Living Legacy

What about the rest of us? Should we care? Could potential oil contamination mean uncertain times ahead for the entire Sound?

"Evaluating oil spills is enhanced by long-term studies that recognize the natural variability of marine environments," states Exxon researcher John A. Wiens, stressing the need for objectivity "divorced from advocacy."

But like oil under rocks, questions persist. For instance, long before the 1989 spill, harbor seals and some sea birds had begun a steep decline. Why? Since the spill, some species have either drastically changed habits or dwindled in numbers. Is the spill solely to blame? Rosenberg, the wildlife biologist, says it's difficult to isolate spill

Valdez-based tour operator Stan Stephens pauses at Growler Island, in northern Prince William Sound. Stephens is board president of the Sound's Regional Citizens' Advisory Council, a watchdog panel that grew out of the oil spill in 1989. (Alissa Crandall)

ABOVE: *Alaska's oil spill trustees set aside $55 million to continue a habitat protection program into the next century. Arnica blooms at Pauls Lake, near Afognak Island's northeast coast. (Roy Corral)*

LEFT: *An exception to the region's rocky shore is a marshy Bay of Isles site in western Prince William Sound where this oily pool was found in 1998. Exxon reported in 1999 that the heavily oiled site displayed a dramatic decrease in weathered oil; critics continue to label it "Death Marsh." (Beth Whitman)*

effects from other ecological change that, he says, could affect "everything."

"There seem to be declines in lots of species throughout the oceans of the world," Rosenberg notes. "It's difficult to get a biological cause and effect. Still, there is evidence that birds and mammals are being exposed to hydrocarbons."

Anecdotal evidence of unexplained change in the Sound comes from residents, fishermen and some scientists. Rosenberg says thousands of scoters — ducks that winter along inshore marine waters — used to gather near Tatitlek Narrows, which border Bligh Island in eastern Prince William Sound.

"Now you can hardly find one during the herring spawn," he says, adding it's unclear whether scoters congregate elsewhere in the Sound or have seen a population decline. Like other researchers, he's trying to find the answer.

But even as the spill zone slowly recovers, a question that surfaced in the cleanup operation's earliest days — how clean is clean? — remains. Today, as in 1989, the answer depends on whom you ask.

"There is no evidence that the biota are stressed by remnants of oil on any shoreline," says Bowdoin College chemistry professor David S. Page in a report prepared for Exxon and released in 1999. Flora and fauna appear normal, based on the type of shoreline, and all locations saw an "abundance" of newly set young, Page concluded.

Robert Spies, a California-based researcher and chief science adviser to the spill trustees since 1989, notes that while Exxon may not object to the Sound's remaining oil, others — such as Alaska's fishing fleet accustomed to pollution-free waters, kayakers seeking out wild places and Alaska Natives who turn to Prince William Sound for food — might.

Oil remaining in the environment "is having some effects on wildlife," Spies said. "But we have little definitive that we can say about the ecological significance of these effects, except that they are not very obvious but might be insidious."

It is this long-range, subtle effect that scientists continue to investigate while holding out hope that, if another big spill occurs, science will have more answers — a living legacy of Alaska's attempt to restore a marine ecosystem. "That," Spies says, "has implications way beyond the spill."

RIGHT: *Machines like this "omni-boom," washing a Knight Island beach, were mobilized at remote oiled sites in 1989. Cleanup costs eventually hit $2 billion. (Patrick J. Endres)*

BELOW: *Revisited in 1990, the same beach shows results of human efforts as well as scouring by winter storms. (Gene Pavia / Alaska Department of Environmental Conservation)*

Here is a closer look at some of the government science undertaken since the Exxon Valdez oil spill. Alaska's inventory of species and resources injured by the accident was prepared by the oil spill trustees and has 26 entries in all.

Bald Eagles

An estimated 5,000 to 6,000 bald eagles inhabited Prince William Sound at the time of the Exxon Valdez oil spill while as many as 10,000 could be found in the wider spill area, which includes coastline outside the Sound. A three-year study determined that eagle population numbers declined because of the spill, but had rebounded by 1995.

The bald eagle, whose numbers in Alaska top all other state's combined, was the only species on the Exxon Valdez oil spill trustees' "recovered" list until river otters were added in February 1999. Cleanup crews found the carcasses of 151 eagles, leading officials to estimate that about 250 bald eagles, or about 5 percent of the population, were killed by oil in Prince William Sound. Scientists believe several hundred more probably died outside the Sound.

Reproductive failure in 1989 was directly related to shoreline oiling near nests, government researchers say. But scientists couldn't determine whether commotion from shoreline cleanup operations caused some birds to abandon nests or if other factors were responsible. Eagles also may have died after eating oiled prey or succumbing to exposure if feathers were oiled, depriving the bird of insulation. Other injuries may have taken a toll as well.

Reproductive success in 1990 apparently was typical within Prince William Sound and in the wider spill zone as well. U.S. Fish & Wildlife Service researchers Timothy D. Bowman, Philip F. Schempf and Jeffrey A. Bernatowicz offered several possibilities for lack of reproductive problems among eagles in the spill zone beyond Prince William Sound. Factors include the later arrival of oil during the nesting season as well as decreased toxicity because oil had weathered.

Scientists radio-tagged 62 eagles four to five months after the spill and another 97 eagles over the next two years. High survival rates were reported among the birds. Scientists then looked at bald eagles from oiled and unoiled areas and compared survival rates based on whether the eagle was tagged in eastern or western Prince William Sound; presence or absence of oil at the capture site; and percentage of time an eagle had spent in oiled areas after tagging.

Scientists found no difference in survival for any of these comparisons.

Scientists believe that before the spill, eagle populations in Prince William Sound increased at a rate of about 2 percent annually. They also estimated the eagle population in Prince William Sound would require about four years to return to the level it would have been in 1989, had the spill not occurred.

An aerial survey in 1995 indicated the eagle population had indeed returned to its estimated pre-spill size. Bald eagles were placed on the recovered species list in 1996.

Harlequin Ducks

Harlequin ducks live most of their lives at sea but migrate inland to nest along whitewater streams. These sea ducks live 10 years or more and do not breed until they are at least 2 years old. If eggs are lost to a predator, for example, no more are laid that year. Changes in population size are strongly influenced by the survival of breeding females.

Research team members Ellie Mather and Kim Trust remove harlequin ducks captured in a mist net. The birds are examined for the presence of a telltale enzyme that, government scientists say, is produced in harlequins exposed to contaminants. (Patrick J. Endres)

Scientists studying the harlequin's post-spill numbers say Prince William Sound ducks from oiled sites display a lower growth rate than harlequins from unoiled areas. To learn why, researchers captured harlequin ducks over a period of years, attached identification bands to their legs and outfitted some with radio transmitters to track their range. Blood and liver samples were collected to detect exposure to contamination based on the presence of the enzyme cytochrome P4501A.

Scientists discovered that adult female survival was lower in oiled areas than in unoiled areas. They also found that birds from oiled areas several years after the spill had much higher P4501A levels than harlequins from unoiled sites. Researchers, who believe Exxon Valdez oil is the most likely contamination source, note that oil may play an important role in population trends of harlequin ducks.

Food does not appear to be the limiting factor. For instance, on an oiled site on Knight Island, more prey was available for each duck than could be found at an unoiled site on Montague Island. Body mass and blood chemistry also appeared similar between oiled and unoiled sites.

If female survival rates at oiled sites continue to be depressed, populations will suffer, researchers now conclude. "Poor female survival equals population declines," said Daniel Esler, a biologist with the Biological Resources Division of the U.S. Geological Survey.

"Populations of animals with life histories like 'harlies' have low growth potential," Esler notes. "That, coupled with less than ideal demographic characteristics (such as low survival) suggests that full recovery from the oil spill will be a long time coming."

A related three-year study by state biologist Dan Rosenberg found that harlequin duck populations have declined significantly in oiled areas while increasing in oil-free regions. Scientists note that while harlequins in oiled and unoiled areas are producing the same number of young, survival rates are lower in oiled areas.

Killer Whales

Prince William Sound is home to both resident and transient groups of killer whales which do not appear to ever intermingle. Resident killer whales travel in stable social groups, called pods, of eight or more. Transient killer whales travel in groups of less than seven and do not maintain a stable social structure. Scientists identify individual whales by markings on dorsal fins.

Prior to the spill, 110 killer whales in six groups regularly traversed Prince William Sound in summer; experts now say the Kenai Fiords area is more heavily used.

One of the Prince William Sound groupings, known as the AB Pod, has undergone significant change since 1989. Before the oil spill, the AB Pod was the most commonly seen killer whale group in the Sound. Unusually friendly, these whales often fearlessly approached and followed research vessels.

Six days after the oil spill, scientists observed the AB Pod at the south end of Knight Island in southwestern Prince William Sound where the pod swam through oil slicks. Seven of the pod's 36 members eventually turned up missing and were confirmed dead.

Missing whales included three adult females and four juveniles. Two of the females left offspring younger than 4 years old. By the spring of 1990, another six whales from AB Pod had disappeared and were presumed dead. These whales included one female that left a young calf, four juveniles and one male. In the following years, three orphaned calves also died.

Government scientists say this unprecedented mortality is "circumstantially" related to the oil spill, since there were no unusual deaths in any of the other well-

known resident pods at the time of the spill or the following year. The dorsal fins of two adult males in the AB Pod folded over after the oil spill, authorities said, and both eventually died in 1991. A folded dorsal fin may be a sign of poor health

While some experts are not convinced that whales were exposed to high enough levels of contaminants to induce illness, others note that inhaling petroleum or petroleum vapors can cause pneumonia, nerve damage and other disorders, even sudden death.

Since the oil spill, another commonly seen pod of killer whales — the AT1 Pod — has lost 11 of 22 members; they were last seen

in 1990. At least one was found dead and three others were seen near the Exxon Valdez shortly after the spill.

A killer whale is presumed dead after it disappears from its pod and is not seen for two consecutive years. No resident killer whale has ever returned to its maternal group after disappearing that long. Scientists believe killer whale carcasses sink because they are rarely found.

Transient whales are harder to track since they sometimes leave their group to join others. Such is the case with the AT1 group, whose members often branch off to travel in small numbers. The AT1 group of transients are unusual because, prior to the

Scientists monitoring Prince William Sound's killer whales after the spill look for clues such as an upright or folded-over dorsal fin; the latter can indicate sickness. Biologist Craig Matkin snaps photos of dorsal fins and "saddle" patches, used to identify individual whales. Studies are ongoing. (Chlaus Lotscher)

spill, they were sighted in Prince William Sound year after year. Scientists believe members of this group either died from

BELOW: *To collect tissue and blubber samples, killer whales are darted with a gun as they surface. The sharp, hollow dart, which draws no reaction from the whales, falls off and floats for retrieval. (Chlaus Lotscher)*

BELOW, RIGHT: *An unusually large group of river otters collects in Nuka Bay within Kenai Fjords National Park, one of two national parks marred by oil in 1989. River otters in 1999 joined bald eagles on a list of species recovered from population setbacks. (Roy Corral)*

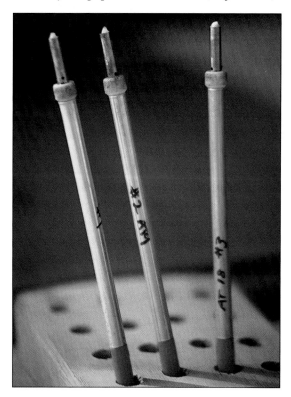

inhaling oil or oil vapors, or from feeding on heavily oiled harbor seals. There have been no new calves in the AT1 pod since 1984, researchers say. Scientists also found contaminant levels in the AT1 group averaged 10 times higher than in resident killer whales.

As for resident killer whales, their overall number in Prince William Sound has increased since the spill except for the AB pod, prompting scientists to conclude that recovery may not be related to ecological changes. Researchers also say they cannot predict if the AB Pod ever will rebound to pre-1989 numbers.

"AB Pod's recovery is severely impeded by the loss of the reproductive females and juveniles that died at the time of the spill," note researchers Craig Matkin of Homer and Fairbanks-based Eva Saulitis, both with the North Gulf Oceanic Society. "This loss may have significantly reduced the

potential production of new calves."

Disruption of the pod's social structure also may hinder recovery. Killer whale pods consist of related maternal groups and their offspring of either sex. Because killer whales are faithful to their pods for life, these social bonds are considered among the strongest of any mammals.

Scientists continue to monitor killer whales annually, keeping track of individuals and collecting skin and blubber samples for genetic and contaminant analysis. Studies show that contaminant concentrations, passed from mother to offspring in milk, vary between resident and transient whales and also among individual whales.

River Otters

Declared a recovered species in 1999, river otters are especially vulnerable to marine pollution because they feed from the intertidal zone, eating a wide variety

of fish and invertebrates. Twelve river otter carcasses were found following the spill but exact losses remain unknown.

Prince William Sound river otters live along the coast. Oil spilled in 1989 covered some of the animal's prime habitat and diminished the diversity of its diet. In 1990, river otters from oiled areas were found to have lower body mass, or reserves, than otters in unoiled areas; by 1992, body mass had improved.

Scientists considered whether lower body weights could be traced to exposure to oil or changes in the amount or vulnerability of prey. Research also examined any change in the otters' choice of habitat to avoid contaminated shoreline. Exposure to oil now seems the likely culprit, government researchers say.

River otters showed differences in size of home-range, diet and blood chemistry in oiled and non-oiled areas following the spill. Government scientists believe the otters also may have seen changes in reproduction and survival, because they moved greater distances and had lower body mass.

Ongoing research by R. Terry Bowyer, J. Ward Testa and James B. Faro has uncovered improvement. "We were uncertain for a long, long time," said Bowyer, professor of wildlife ecology at the Institute of Arctic Biology, University of Alaska Fairbanks. "All the recent data suggests recovery, if not complete, will be shortly."

Sea Otters

Unique among marine mammals, sea otters are not insulated against the cold Pacific Ocean with a layer of fat. Unlike blubber-possessing whales, dolphins and seals, sea otters depend on their dense fur to

trap tiny air bubbles and keep out the chill.

To stay warm, sea otters also must maintain a high metabolic rate, eating the equivalent of about 25 percent of their body weight every day. Sea otters typically consume from 10 to 25 pounds of food a day, primarily clams, crabs, urchins and mussels and other invertebrates in shallow coastal waters.

In 1995, scientists counted 13,000 sea

A sea otter passes time in Valdez where oiled otters were cleaned and monitored at a rescue center before release. (Alissa Crandall)

otters in the Sound but the exact pre-spill population is unknown. The lack of data has hindered authorities trying to determine when sea otter numbers can be said to have rebounded.

Several thousand sea otters died after being exposed to oil from the Exxon Valdez. Although the die-off was modest when the overall Alaska sea otter population was taken into account, sea otters in the most heavily oiled areas of Prince William Sound still have not returned to prespill numbers. Scientists note that findings are specific to certain, heavily oiled sites and may not apply to the wider spill zone.

One model used by scientists cautiously suggests it might take 10 to 23 years before the sea otter population in oiled areas regains pre-spill abundance. The projection is deemed a best-guess forecast, researchers advise, because assumptions that may not all come to pass were used in the information-gathering process and in the model.

Sea otters were vulnerable not only

because oil-matted fur failed to provide insulation. The otters' peak pup season occurs in late spring so that many females were pregnant or nursing when exposed to oil. The sea otter, whose gaze seems to lock on people, became a symbol of wildlife losses after the spill and rehabilitation centers were quickly set up in Valdez and Seward. More than 300 sea otters were caught in Prince William Sound. Many were heavily

Efforts to aid wildlife were under way within a week of the tanker accident. An oiled sea otter captured on Crafton Island in Prince William Sound (facing page) is transferred to a helicopter for a flight to Valdez where fur is sudsed (above) at a rescue center. Otters were then kept in holding pens to monitor recovery. Some were released to the wild while others went to sites (right) such as Sea World of San Diego, Calif. (All: Alissa Crandall)

oiled and did not survive. About 200 otters did live through oiling, capture, cleaning and captivity to be released into the wild in late summer 1989. Monitoring showed that survival and reproduction rates following release were relatively poor.

"This (low survival rate), combined with the high cost of operating the treatment centers, and recognition that the stress of capture and holding was in itself harmful to the otters, made the value of the rehabilitation process somewhat controversial," note researchers J.L. Bodkin and B.E. Ballachey of the U.S. Geological Survey's Biological Resources Division.

Nearly 1,000 sea otter carcasses were recovered in oiled areas after the spill, including 500 from Prince William Sound. Many of these were found to have damaged lungs, livers and kidneys. Between 1990 and 1993, researchers found 14 percent more otters in unoiled areas of Prince William Sound and 35 percent fewer otters in oiled areas, compared to pre-spill counts.

In 1990-91, mortality of prime-age sea otters was abnormally high in oiled areas, government studies show. However other studies suggested that reproduction and survival of adult female sea otters in the western Sound was normal. Survival of pups in the western Sound in the winters of 1990-91 and 1992-93 was poor and lower than in eastern Prince William Sound, an

oil-free area. While skiff surveys over a five-year period ending in 1996 indicate the sea otter population in oiled areas of Prince William Sound had not increased since the spill, more recent aerial surveys suggest some improvement. Data from the Sound's oiled western zone, collected from 1993 to 1998, indicate a 10 percent annual increase in sea otter abundance; however no increase was detected in the aerial survey in worst-case portions of northern Knight Island.

Scientists are continuing to monitor sea otters in oiled areas through the Nearshore Vertebrate Predator Study, a restoration project begun in 1995. Findings could help gauge the sea otters' progress while determining if oil exposure has long-term effects among the animals.

Injured Species and Resources at a Glance

EDITOR'S NOTE: *This list is based on research released in 1999 by the Exxon Valdez Oil Spill Trustee Council and Exxon Corp.*

Exxon and the spill trustees disagree on the extent of recovery in the spill zone, and the company has appealed a $5 billion punitive judgment stemming from an oil spill class action.

ALASKA GEOGRAPHIC® has not attempted to reconcile discrepancies over recovery; this list is presented as a summary of research.

ARCHAEOLOGICAL RESOURCES: Recovering, based on preservation in progress and low rates of spill-related vandalism. More than 3,000 archaeological or historically significant sites may be included within the spill zone. *Exxon: Any lingering effect of the spill is negligible or undetectable; disagrees that recovery is not yet achieved.*

BALD EAGLES: Classified as fully recovered in 1996 based on productivity and population surveys. *Exxon: Sustained impact, recovered.*

BLACK OYSTERCATCHERS: Recovery is under way. No oil-related differences noted in clutch size, egg volume or chick growth rates. *Exxon: Sustained impact, recovered.*

CLAMS: Spill impacts varied by species, degree of oiling and location. Data indicate that littleneck clams were killed and displayed slower growth rates after the spill; a 1993 study showed no significant differences in hydrocarbon concentrations between littlenecks in oiled and unoiled areas. Clams are classified as recovering. *Exxon: Impacted by oiled shoreline and cleanup treatment. Current status reports should take into account heightened predation by sea otters.*

COMMON MURRES: Post-spill monitoring at breeding colonies indicates normal reproductive success has been sustained since 1993. Recovery is under way. *Exxon: Acute loss followed by recovery.*

CORMORANTS: Pelagic, red-faced and double-crested cormorants were among bird carcasses found after the spill. Cormorants are considered not recovering based on surveys through 1998. *Exxon: Pelagic cormorant sustained impact, recovered.*

CUTTHROAT TROUT: Recovery status unknown pending the outcome of 1999 studies including growth rate based on geographic location. *Exxon: No impact.*

DESIGNATED WILDERNESS: Oil was deposited above the mean high tide line at eight areas designated as wilderness or wilderness study areas including Katmai National Park, Chugach National Forest, Kenai Fjords National Park and Kachemak Bay Wilderness State Park. Recovery status is unknown pending surveys in 1999. *Exxon: Vast majority of shorelines were recovered within two or three years of the accident.*

DOLLY VARDEN: Recovery unknown pending the outcome of genetics and growth data. Preliminary results suggest recovery from initial growth-related effects is likely. *Exxon: No impact.*

BELOW: *Barren Island breeding colonies saw normal reproductive success among murres by 1993; El Nino ocean warming of 1998 apparently caused some disruption. A thick-billed murre shelters its egg. (Patrick J. Endres)*

RIGHT, TOP: *Results of egg mortality studies are among indicators that pink salmon are recovering from oil spill losses. (Alaska Department of Fish and Game)*

RIGHT, BOTTOM: *After the spill, numbers of spawning herring were depressed in Prince William Sound through 1995; sharp increases were detected in 1997 and 1998. (Ruth Fairall)*

HARLEQUIN DUCKS: Not recovered based in part on enzyme studies in 1998 revealing exposure to hydrocarbons. *Exxon: Sustained impact, recovered.*

HARBOR SEALS: Gulf of Alaska populations were declining before the oil spill, which marred key haul-out areas and killed about 300 seals outright. Not recovered, based on an average estimated population decline of about 5 percent annually. *Exxon: More likely that "missing" seals were displaced by disturbance of heavy cleanup and not lost from the Sound's population. Long-term population decline is unrelated to the spill.*

INTERTIDAL COMMUNITIES: Recovery in many locations is substantial. The common seaweed *fucus gardneri*, which provides cover for invertebrates, has yet to fully recover on direct-sunlight shores.

Exxon: *Some communities damaged in short term by cleanup, including hot water washing. Recovered.*

KITTLITZ'S MURRELETS: Found only in Alaska and portions of the Russian Far East, little is known about these murrelets which favor tidewater glaciers; 72 carcasses were found after the spill. Full impacts of the spill and recovery status are unknown.
Exxon: *Notes that murrelet favors heads of glaciers, a region beyond the spill area. No conclusion as to impact or recovery.*

KILLER WHALES: Not recovering based on lack of sustained stability of pods or recruitment into them.
Exxon: *No spill impact noted; questions "intrusive" techniques of agency studies that could prompt pods to relocate.*

LOONS: Four species of loons were represented in carcasses collected after the spill; most were common loons that probably included a mix of resident and migrant birds. Population surveys through 1998 show no sign of recovery.
Exxon: *As a group, recovered.*

MARBLED MURRELETS: Productivity based on surveys of adults and juveniles on the water appears within normal bounds; a population increase in oiled areas was noted between 1990 and 1993. Recovering.
Exxon: *Population is stable.*

LEFT, TOP: *River otters were declared a recovered species based in part on habitat-use comparisons in oiled and oil-free areas. (Exxon Valdez Oil Spill Trustees)*

LEFT, BOTTOM: *Long-lived and slow to reproduce, common loons accounted for about half of loon carcasses found after the spill. (U.S. Fish & Wildlife Service)*

BELOW: *Black oystercatchers number between 1,500 and 2,000 in southcentral Alaska and occupy intertidal zones. (James L. Davis)*

MUSSELS: Recovering. Significant natural declines in oil suggest a return to background concentrations in next few years.
Exxon: *Studies through 1998 reveal a "dramatic" decline in oil in mussel bed sediments.*

PACIFIC HERRING: Recovering. Increased biomass noted in 1997 and 1998.
Exxon: *Complex species with pre-spill population crashes in Prince William Sound. No oil spill impact.*

PIGEON GUILLEMOTS: In decline before the spill, possibly due to reduced availability or quality of prey. Not recovered, based on surveys of oiled shorelines showing depressed pigeon guillemot numbers .
Exxon: *Densities in both oiled and oil-free sites showed significant increases from 1989 to 1998. Regional changes rather than localized spill effects may be a factor.*

PINK SALMON: Recovering, based in part on egg mortality studies.
Exxon: *No obvious detrimental effects; five largest harvests recorded since the spill. Questions protocol of egg-mortality studies.*

RIVER OTTERS: No indications of lingering injury. Recovered.
Exxon: *Concurs.*

ROCKFISH: No recovery objective identified since original extent of any population loss is unknown.
Exxon: *Did not study.*

SEA OTTERS: A long-lived mammal with a low reproductive rate, sea otters are classified as recovering in much of the spill zone except for heavily oiled sites in the western Sound. Food supply and continuing exposure to hydro-

carbons are being studied as possible hindrances.
Exxon: *Impacts, followed by clear evidence of subsequent recovery.*

SEDIMENTS: Recovering, based on 1993 evidence that little Exxon Valdez oil and related higher levels of microbial activity were found at most test sites. "Substantial" subtidal oiling persists at Herring, Northwest and Sleepy bays, all in the western Sound.
Exxon: *Scattered areas of oil remain, notably at Sleepy Bay; oil residue is not biologically active.*

SOCKEYE SALMON: Recovering. Overescapement into the Kenai River following fishery closings in 1989 prompted concern that juveniles would overgraze zooplankton, eventually resulting in lost sockeye production. Negative effects were apparent from the brood years 1989 to 1992; returns from 1993 to 1995 are not complete because some fish are still at sea but results to date suggest a return to normal levels. Populations at both Red and Akalura lakes, on Kodiak Island, also are recovering from overescapement.
Exxon: *Negative effects do not take into account large escapements in 1987 and 1988 which can lead to "cyclic dominance," in which a big run is followed by smaller runs. Overescapements are the result of management decisions, not oiling.*

SUBTIDAL COMMUNITIES: Recovering, based on the status of eelgrass stands which support invertebrate life such as worms, snails, clams and sea urchins.
Exxon: *Findings are "extremely" site specific. Comparisons done in 1990 and 1991 at Bay of Isles, an oiled site, and Drier Bay, an oil-free site, both on Knight Island, suggest findings on subtidal communities cannot be applied baywide.* •

Harbor Seals

Harbor seals are among the most common marine mammals in Prince William Sound, where they live year round. While the exact harbor seal population strength

Diet, disease and predators could provide clues into the harbor seals' decline in Prince William Sound — a drop that began 20 years ago. These seals seek shelter in Unakwik Inlet, along the Sound's north shore. (Alissa Crandall)

is unknown, a decline ranging to as much as 80 percent over the past 20 years has been detected. The reason remains a mystery.

Seals came into direct contact with Exxon Valdez oil in water and on land. The animals swam through oil and breathed in vapors at the surface. Hauled out on shore, harbor seals crawled through oil and rested on oiled rocks and algae.

In the most severely oiled section of central Prince William Sound, more than 80 percent of seals observed in May 1989 were

oiled; abnormal behavior was noted a few months later. Oiled seals appeared sick, lethargic or unusually tame. They exhibited excessive eye tearing, squinting and disorientation.

In the first few months after the Exxon Valdez oil spill, 18 harbor seals were found dead or died in captivity. Fifteen seals had oil on their bodies and three were pups. Four seals suffered bleeding of internal organs, two had severe skin irritation, two had inflamed eyes and three had symptoms

of malnutrition. Three seals showed evidence of brain damage.

Scientists estimate that about 300 seals died in Prince William Sound due to the oil spill. By early September, five months after the wreck, obvious effects of oiling on harbor seals had vanished. Most seals older than pups had molted, shedding their oil-stained hair. They did not seem to be reoiled since much of the slick had dispersed and most places where seals hauled out had been cleaned.

One year after the spill, however, harbor seals from oiled areas in Prince William Sound had a higher level of oil-related hydrocarbons in bile compared with seals in unoiled areas, government research shows. Researchers believe the animals still were being exposed to oil. From 1990 to 1995, population trends seemed similar in oiled and unoiled areas yet the harbor seal population continued to decline at about 6 percent a year in both zones.

To learn why, scientists are studying the seals' diet, disease and predators; researchers also are placing satellite tags on harbor seals to learn about their movements and diving behavior.

Kathy Frost, an Alaska Department of Fish and Game biologist in Fairbanks, notes that while losses blamed on the oil spill contributed to the species' ongoing decline, harbor seal numbers were dropping and continue to decline today without a clear explanation. So far there is no indication that disease is to blame.

Mussels

Many mussel beds in the spill area, particularly those on soft sediment, were untouched immediately after the accident to

Researchers flag a bed of dead mussels to monitor restoration efforts in the Sound. (Cary Anderson)

give natural cleansing a chance. Authorities hoped turbulent weather — winds, waves and storms — would remove oil from mussels and underlying sediments in the beds, avoiding damage from high-pressure hot water hoses that had been aimed at oiled shorelines. While effective in places, hosing had destroyed much of the marine life in intertidal zones, scientists said.

But two years after the spill, oil remained in sediments of untreated mussel beds. Embedded oil was a potential source of contamination for mussel consumers including sea otters, birds and people, as well as for animals that live in the mussel beds — littorine snails, barnacles and marine worms. It also appeared that oil from trapped sediments continued to recontaminate mussels as it dispersed into water surrounding the beds.

While oil contamination in some beds declined naturally from 1992 to 1995, government scientists concluded by mid-1993 that mussels had yet to fully recover from the spill. Three years later, mussel samples revealed elevated levels of petroleum hydrocarbons in tissues. Effects of lengthy exposure to oil on mussels were hard to demonstrate. Even in the most contaminated beds, mussels grew and reproduced. Studies found only some cell

One year after the tanker wreck, mussels and other marine life revive in Prince William Sound. (Cary Anderson)

abnormalities and reduced tolerance to the air in mussels from oiled beaches. Still, mussels were thought to be a pathway for oil to other wildlife since creatures that fed on mussels showed evidence of continued oil exposure

Scientists began seeking ways to reduce oil in the beds, where mussel clumps form a kind of armor over underlying sediment. In one experiment, a strip of mussels was removed to increase natural flushing of the bed. The result: oil concentration was not significantly reduced and mussels recolonized the strip within three months. In another experiment, researchers transplanted several small patches of mussels from two oiled beds onto clean sediment. Transplanted mussels cleansed themselves of oil quickly, but the die-off was high.

Scientists then tried manually restoring the beds. They removed oiled mussels and replaced contaminated sediment with clean before returning mussels to the bed. Again, oil concentrations quickly lowered but not always significantly. There also was some evidence that sediments were contaminated again from deeper or surrounding sediment. Die-off varied and the method was considered a partial success.

Stan Rice and Mark Carls, researchers at Auke Bay Laboratory within the National Marine Fisheries Service, suggest another restoration technique. Mussels could be removed from contaminated soft gravel or mud and their shells cleaned for external oil; mussels would then be placed in floating pens where they could rid themselves of internal oil. Back at the beds, extensive cleaning could be undertaken using backhoes, hot water washes or chemicals and fertilizer to promote the natural breakdown of oil.

Mussels could be returned to cleansed sites where they would reattach, stabilize the habitat and once again become a source of food in the Sound. Auke Bay researchers plan to survey oiled mussel beds again in 1999. The scientists recommend monitoring mussel beds every three years, until oil concentrations in sediments and the mussels themselves are at pre-spill levels.

Pink Salmon

Four to five months after the Exxon Valdez oil spill, pink salmon returned to spawn in streams that cut through oiled beaches. government scientists say incubation downstream from oiled beaches decreased survival and impaired reproductive strength for at least four years after the spill.

Although oil may not have entered the streams themselves, it contaminated gravel surrounding the streams and, researchers believed, may have contaminated pink salmon eggs. To test this observation, egg mortality rates were monitored in oiled

Ongoing studies of pink salmon eggs show oil has harmful effects at lower pollution rates than once thought, government scientists say. They have classified the fish as "recovering" in the spill zone. Exxon research concludes that pink salmon runs endured no obvious detrimental effects. Here pink salmon churn a stream near Valdez. (James L. Davis)

and oil-free streams; fry that survived incubation in oiled gravel were then studied to see if oil caused long-term damage.

Pink salmon have a two-year life cycle: After spawning in intertidal areas in August and September, eggs develop in gravel, hatch in midwinter, and remain in gravel until the yolk is completely absorbed, usually in early spring. Fry are then ready to emigrate to sea.

Scientists selected 31 streams to monitor and divided them into four zones, three intertidal and one above the point where oil made its highest landfall, an area called the "bathtub ring." In 1989, embryo mortality was highest in oiled streams in the intertidal zones but there was no difference above the bathtub ring.

In 1990, embryo mortality was highest only in the zone that contained the bathtub ring. In 1991, when the 1989 fry returned, embryo mortality was high in all zones of the oiled streams, even sections that had escaped contamination. The occurrence repeated itself, but was weaker, when the 1990 fry returned in 1992. This led scientists to believe that adult fish returning to oiled streams had an impaired reproductive ability linked to oil exposure during incubation.

Scientists tested this theory by raising eggs from oiled and unoiled streams. Hatchery studies showed that differences in survival stemmed from parents, and not the stream itself. This means that high egg mortality seen above the bathtub ring in 1991 could have resulted from damage to the reproductive health of the 1989 brood. Government cientists determined oil's harmful effects were occurring even at very low concentrations — findings that had not been noted before.

Research is ongoing. When scientists incubated salmon eggs in gravel contaminated with known quantities of oil, they confirmed that oil caused egg mortality and found it affected long-term fish growth and survival. Researchers Ron Heintz, Stan Rice and Jeffrey W. Short, all of the National Marine Fisheries Service, and Brian Bue and James Seeb, with the Alaska Department of Fish and Game, summed up their findings in a report to the oil spill trustees: "Growth was significantly lower in fish exposed to oil with polycyclic aromatic hydrocarbon (PAH) concentrations as low as 1 part per billion in water, which is one order of magnitude lower than the Alaska State water quality standard for PAH (15 ppb in water)." Polycyclic aromatic hydrocarbons are the most toxic class of compounds contained in crude oil.

Pacific Herring

Few species are of greater ecological and economic importance in Prince William Sound's food chain than the Pacific herring. The tiny fish is sought out by other fish, humpback whales, harbor seals and marine and shore birds as well as jellyfish and other invertebrates.

Herring and their eggs also provide a multimillion-dollar resource for commercial fishermen in the spring. Bony and fast-moving, herring are about 8 inches long, travel in schools and have very oily flesh. Layers of crystals in herring skin reflect light and serve as camouflage.

Pacific herring populations vary tremendously depending on predators, larvae survival, disease, available food, fishing pressure and exposure to oil. Before the Exxon Valdez spill, Prince William Sound's herring population was high and increasing. Estimates of the biomass of spawning adult herring ranged from a low of 16,400 metric tons in 1994 to a high of 113,200 metric tons in 1989. (Biomass estimates express a population size by weight.) But the 1989 year class of herring failed to join the spawning population in 1993 and only about 25 percent of the forecasted return actually showed up. The herring fishery was curtailed, then closed from 1994-96. A limited commercial harvest was allowed in 1997 and again in 1998.

Prince William Sound is a vast laboratory for Pacific herring researchers. Studies show that a herring fishery crash in 1993 cannot be blamed solely on spilled oil in 1989. (Roy Corral)

The oil spill in 1989 occurred just a few weeks before herring spawned in Prince William Sound. The fishery closed that year because many herring were gathering in oiled waters to spawn. About half of the egg biomass was deposited within reach of spilled oil and an estimated 40 percent to 50 percent was exposed to oil during early development.

Adult herring also were exposed to spilled oil but effects are not clear: Adults examined immediately after the spill had liver lesions attributed to oil. Studies show lesions may have been caused by a viral disease that, in turn, might have been triggered by oil exposure.

Scientists discovered physical abnormalities and genetic effects when herring eggs were exposed to oil. Evelyn Brown, a University of Alaska Fairbanks scientist, and Mark G. Carls with the National Marine Fisheries Service, noted that results observed in the laboratory were identical to what happened to herring larvae of different ages from oiled areas within Prince William Sound during spring 1989. Effects included premature hatching, low larval weights, reduced growth, increase in physical deformities, population decline and genetic damage.

When commercial salmon fishing was closed in parts of the spill zone, too many spawning fish returned to Kodiak lakes and the Kenai River. Managers worried that juvenile sockeye would "overgraze" food, eventually resulting in a population decline. Negative effects were noted in the Kenai River; trustees now say more normal sockeye returns are likely. (Patrick J. Endres)

While equal biomass of eggs was deposited in oiled and unoiled areas, it is estimated that oiled areas produced only 17 million viable larvae compared to 12 billion from oil-free areas.

Larvae and juveniles also may have eaten contaminated food. Copepods, for example, are a class of large zooplankton that are key prey for herring, salmon and pollock among other fish. Copepods are known to accumulate and concentrate petroleum hydrocarbons in their bodies, potentially passing along contamination. (Studies show that zooplankton may absorb up to one-third of subsurface oil particles after a spill.) Herring contamination may also have come from earlier exposure of incubating eggs.

Following the Flow in Prince William Sound

By Kris Capps

Prince William Sound is a shallow sea surrounded by mountains, fiords and coastal rivers. Depths in the central Sound range from roughly 1,300 feet to 1,400 feet while in the northwest Sound, the "black hole" basin is twice that deep. This basin is a main wintering area for zooplankton, especially one called "zoop" that is a favorite food of juvenile fish. Without that deep spot, there would be no food source for hungry juvenile fish in spring.

To understand the path of Exxon Valdez oil through Alaska waters, and to potentially predict plankton bloom abundance, researchers began by tracing the Sound's link to the Gulf of Alaska. The two bodies connect through Hinchinbrook Entrance and Montague Strait. The flow is generally thought to be inward at Hinchinbrook Entrance and outward at Montague Strait, except when altered by seasonal changes in wind and precipitation.

Scientists were surprised to learn that in summer and fall, currents flow in at Montague Strait and out at Hinchinbrook Entrance. Speed of the current is determined by tides.

Research by scientists Shari L. Vaughan, L.B. Tuttle, K.E. Osgood and S.M. Gray, all of the Cordova-based Prince William Sound Science Center, shows that the Sound is coldest, saltiest and most homogeneous in March and warmest, freshest and most stratified in September. Little is known about seasonal and year-to-year changes in the Sound's water circulation.

Oceanographers found no indication that Gulf of Alaska waters flowed into the central Sound in April or May. In September, in fact, the Sound seems entirely sealed off from the Gulf of Alaska when a counter-clockwise circulation dominates the central Sound.

Oceanographers were able to determine at what depths currents flowed both in and out of the sound. Circulation dynamics seem fairly stable September through March but become more variable in April, May and June — transition months when plankton are blooming. Oceanographers now believe even small changes in current dynamics may affect growth rate and distribution of phytoplankton and zooplankton; circulation can keep plankton in the Sound, distribute it or flush it out completely.

Oceanographers also studied the physical properties of the upper layer of water and how they related to distribution of zooplankton from 1995 to 1997. Differences in the timing and intensity of vertical stratification formation appear to be the primary link.

To determine where Prince William Sound currents flow, oceanographers deployed five drift buoys consisting of a 20-inch-diameter float with a radio antenna mounted on top and, connected to the bottom, a canvas sea anchor that looks like an open-ended cylinder.

A small shock-absorbing buoy was connected in between, allowing the drift buoy to flow with currents at depths of 30 feet to 50 feet without being influenced by surface winds. Electronics radioed the buoy's position to a satellite every hour. Oceanographers inadvertently learned a bit about currents outside the Sound as well when two buoys drifted past the Kenai Peninsula, 200 miles or more to the southwest. •

How cold, how swift, how deep: Prince William Sound's currents are tracked to understand how fluctuations in the ocean itself may foster or displace marine life. (Roy Corral)

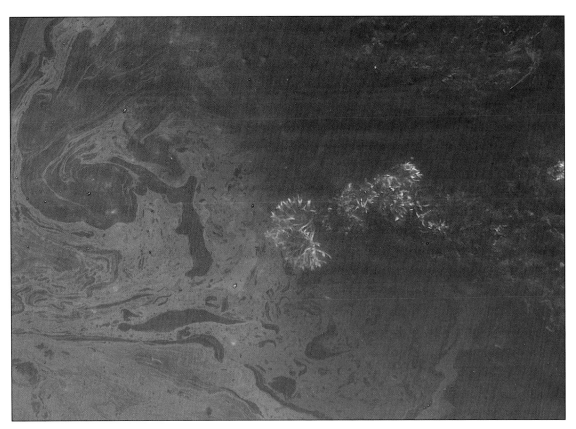

Despite dramatic losses to the 1989 brood year, oil exposure cannot be singled out as the cause of the 1993 herring fishery crash, researchers have concluded. The group was relatively small and plagued by a virus. Other possible causes for the disease might be high population density, food scarcity, poor ocean conditions or previous exposure to oil.

Scientists concerned that oil-exposed herring might suffer genetic damage passed on to future generations so far have uncovered no strong evidence indicating a threat.

What happened to the 1993 herring? One theory is that other species of fish took over the ecological niche once occupied by herring, thus slowing or reducing the likelihood of a population rebound. Pollock apparently increased in numbers, setting up a possible competition for food and pressure on other life stages.

Current research is looking at the possible relationship between disease epidemics and the practice of gathering large numbers of spawning herring into small enclosures, such as in the "pound" fishery.

Findings so far suggest that mortality in the larval and juvenile stages, not the embryo stage, has the greatest effect on the number of herring surviving to adulthood. Since larvae are difficult and expensive to study, studies now focus on the surviving juveniles.

Plankton

Each spring, in April and May, the waters of Prince William Sound turn murky not from sediment but from a population explosion of billions of phytoplankton and zooplankton. These microscopic-sized plants and animals represent the base of the Sound's food chain.

In late March, as Exxon Valdez oil gushed into the Sound, plankton were just beginning that year's population boom. The bloom drifted and reproduced in warmer, sunlit water near the surface where the tiny plants and animals quickly become easy targets for fish, marine birds and mammals. Immediately after the spill, scientists discovered exceptional levels of zooplankton in the water — some of the highest ever recorded. (Because the floating

slick made it hard to sample plankton, scientists used a fire hose to clear a path for their collection nets.)

Oceanographers found no evidence that spilled oil affected plankton. One theory is that any oil-killed plankton were quickly replaced by new plankton coming in with the Alaska coastal current. Prince William Sound is a "flow-through" system; new plankton are continually being flushed into and out of the Sound from the adjacent continental shelf.

After 1992 and 1993, when the Sound's pink salmon and herring numbers fell dramatically, scientists examined whether delayed effects of the oil spill were to blame. Since both juvenile herring and salmon eat zooplankton, oceanographers set out to understand more about how the Sound produces plankton.

They learned that plankton bloom varies depending on ocean conditions. For instance, when surface waters are quiet and calm, phytoplankton bloom is intense but short-lived and zooplankton don't have enough time to repopulate. Much of the plant matter ends up sinking to become food for bottom-dwellers. But when spring weather is windy and cool, the bloom of tiny plants lasts much longer, allowing zooplankton to eat more plants and

Common murres gather on the Barren Islands, off the Kenai Peninsula's southern tip. Carcasses of 30,000 oiled birds were retrieved within four months of the spill; nearly three-fourths were murres. Today the common murre is listed as a recovering species, based on nesting site surveys. (Roy Corral)

Oil spill science is taking a wider view of the ocean and the life it supports by examining food abundance. Plankton, the microscopic plants and animals found in fresh- and saltwater worldwide, are a direct or indirect source of food for all marine animals, including this Steller sea lion. (James L. Davis)

produce more offspring. The same weather conditions, meanwhile, allow some plants to be more easily grazed. These factors explain why, for zooplankton, some years are better than others.

Oceanographers also discovered that zooplankton abundance in spring tends to increase survival rates of small fish such as juvenile salmon and herring because bigger fish will forego small fish in favor of eating plankton. University of Alaska researchers R. Ted Cooney, C. Peter McRoy and David Eslinger have noted that this relationship between zooplankton abundance and small fish survival had not been understood previously. Scientists say that more plankton studies ultimately could aid managers overseeing the Sound's pink salmon and herring fisheries.

A freelance writer who contributes to books, magazines, guidebooks and ALASKA GEOGRAPHIC®, Kris Capps was a reporter for the Fairbanks Daily News-Miner *in 1989 when oil spilled into Prince William Sound, a place she first visited as a fishing boat deckhand. She has since explored the Sound by small plane, sea kayak, sailboat and research vessel. "It is still exquisitely beautiful," Capps says.*

Healing Lands:
Alaska Invests in Nature

By Natalie Phillips

In the heart of Prince William Sound, Jackpot Bay's 100-foot spruce and hemlock tickle a royal blue sky. Salmon streams thread through deep forest favored by marbled murrelets, an elusive fast-flying sea bird. Harlequin ducks and sea otters shelter in Jackpot's tranquil waters, a scene common throughout the Sound where pin-dot islands, serene backwaters and pristine forest are a haven to wildlife.

Isolated Jackpot Bay, 60 miles southwest of Valdez, is among the Sound's prizes, claimed by wildlife, recreational boaters and a private corporation that owned the land and hoped someday to develop it, possibly for logging. Today, thanks to riches that came to Alaska following the Exxon Valdez

FACING PAGE: *Prince William Sound's proximity to population centers in Anchorage, Valdez and Kenai is a boon to boaters seeking relatively serene waters. Kayakers in Harriman Fiord, in western Prince William Sound, are dwarfed by Mount Gilbert decked in summer snow. (Patrick J. Endres)*

oil spill, timber cutting is no longer an option in Jackpot Bay.

In the 10 years since the tanker accident in 1989, nearly 760,000 acres of coastal land, uplands and islands — a combined area roughly the size of Yosemite National Park — has been purchased or protected from development. Money to complete the land deals, some $400 million in all, was spent from a settlement fund paid by Exxon to make amends for the spill. Purchases and conservation easements (legally binding agreements that bar owners from developing their land) spread from rocky Bligh Reef where the Exxon Valdez went aground to Kodiak Island's southern tip, some 450 miles away.

Roughly 1,400 miles of Alaska shoreline were purchased — the equivalent of Oregon's coastline — as well as 280 salmon spawning streams, a half-dozen islands and numerous inlets and bays, including Jackpot. The goal: To provide protected, oil-free nesting, molting and feeding sites for the two dozen species of marine mammals and sea birds whose numbers were cut by the spill. Tracts purchased throughout southcentral Alaska have added substantially to existing public lands,

precisely the legacy that local and national environmental lobbies hoped for.

Matt Zencey, manager of the Anchorage-based Alaska Rainforest Campaign, says trustees who oversee the settlement fund and guided wildlife habitat purchases recognized that nature will heal itself, if given a chance. "The human ability to manipulate the environment and make it recover from injury is pretty limited," he said. In Prince William Sound and on Afognak and Kodiak islands, the Exxon Valdez Oil Spill Trustee Council also has secured agreements to bar clearcut logging.

But no legacy goes unquestioned. In Prince William Sound communities displaced by the spill and the four-year cleanup, townspeople favored spending on science that would help them better understand the intricate relationship among sea life that, in some cases, has fed families for generations. U.S. Sen. Frank Murkowski, R-Alaska, has challenged trustee spending that transformed privately held land — prime for development — into additional forest, park or refuge lands.

In fact, all of the trustee-purchased lands are slated to become new parks or already have been folded into existing tracts

Forested lands held by Chenega Corp. were sold to the trustee council in 1997 after Alaska Native villagers carefully weighed the price of development against the council's $34 million offer to hold acreage for conservation. (Alissa Crandall)

But Randall, who has lived in the bay more than 30 years, rejoiced: "Finally the powers of goodness prevailed." The parcel, purchased with oil spill settlement money, was turned over to the state and in 1994, Afognak Island State Park was established. The destination has begun catching on with sea kayakers, deer hunters and fishermen targeting silver salmon.

Oil spill lawsuits filed by the state and federal governments were resolved out of court in 1991 when Texas-based Exxon agreed to pay $900 million to settle civil lawsuits and another $100 million to settle a criminal pollution complaint. The settlement established the Exxon Valdez Oil Spill Trustee Council, a panel of six top federal and state government officials with authority, on unanimous votes, to spend the bulk of the settlement fund. Their mandate: To fund projects that would restore the spill zone's health.

"Our goal was to protect as much high and moderately ranked habitat as we could," said Deborah Williams, an Anchorage-based trustee who represented the Interior Department. "We achieved most of our goals."

Council members started by drawing a line around the area touched by the spill, which included 1,300 miles of coastline. With the help of government biologists, trustees also identified "injured species,"

including Chugach National Forest in Prince William Sound; Kachemak Bay State Park near Homer; Kodiak National Wildlife Refuge on Kodiak Island; Kenai Fjords National Park near Seward; and Shuyak Island State Park on the north end of the Kodiak Island archipelago.

Land for Sale:
Alaska's Renewal Begins

Late in the afternoon on May 13, 1993, Shannon Randall heard the buzz of chainsaws stop dead outside her Afognak Wilderness Lodge in Seal Bay.

In downtown Anchorage, 300 miles north,

the trustee council had just decided to buy 41,549 acres of logging lands harvested by Seal Bay Timber Co., a subsidiary of Akhiok-Kaguyak Inc., the Native corporation landowners. The trustees' $39.5 million decision saved the land from becoming a clearcut — and stopped the chainsaws instantly.

Among the council's first purchases, the Seal Bay tract is on the far northeast tip of Afognak Island, due south of Shuyak Island. Critics included the Resource Development Council, whose Anchorage office quickly labeled it the most expensive timber on record.

a list of sea birds, fish and marine mammals whose numbers declined in the slick's wake. After devoting several weeks in 1993 to traveling the state, listening to public comment and reading hundreds of pages of testimony, the council's spending plan took shape.

Councilmembers looking back on the decade say they are struck by how closely the panel's early roadmap dovetails with projects completed. Nearly all of the settlement money has been spent or allocated and the trustee council, a fixture in Alaska's oil spill landscape, is nearing the end of its work. Its spending legacy includes:

• a $39 million contribution to help build the Alaska SeaLife Center, a 115,000-square-foot research and marine education site in Seward;

• $150 million for scientific studies including oil's effect on spawning salmon and herring and a collaborative effort by biologists to better understand the Sound's ecosystem;

• $17 million for community-based or small projects such as park visitor centers, new public-use cabins, new hiking trails and boardwalks, new campgrounds and a community museum in Kodiak;

• $243 million for administrative costs, including conducting public hearings, producing public information about the spill and paying legal fees and oil spill cleanup costs as required by the settlement agreement;

• $400 million for the purchase of 760,000 acres of land and conservation easements

New growth flourishes on a Prince William Sound beach. (Al Grillo)

in Prince William Sound, Kachemak and Resurrection bays, and on Kodiak Island and islands nearby;

• $108 million for an endowment to fund future projects, including continued land buys and research.

Molly McCammon, the trustee council's executive director, says that faced with environmental "catastrophe," the panel's priorities quickly became clear. "The first order of business is to ensure that there is not activity on lands that would stress (wildlife) populations while they are recovering," McCammon said.

But deciding which lands to buy and protect was not as easy as drawing circles

on a map. The council developed a detailed ranking system to seek out spill zone lands and other tracts that offered the best habitat for injured species. The ranking matrix evaluated lands for purchase through the filter of species injured and opportunities lost for recreation, commercial fishing or traditional hunting and gathering. Lands that offered shelter to numerous wildlife while also offering fishing, recreation and subsistence opportunities ranked highest.

Nearly all the large tracts of land purchased by the council were held by Alaska Natives through corporations that had been set up under federal law in

1971. The Alaska Native Claims Settlement Act resolved the claims of Tlingit and Athabaskan Indians, and Yupik and Inupiat Eskimos, among other Native groups, who lost homelands when Alaska was purchased from Russia in 1867.

Under ANCSA, Native groups formed corporations and selected 44 million acres of public land for eventual development, to generate money for shareholders. Selected were some of Alaska's finest timber tracts and salmon streams. Before relinquishing lands to foster oil spill restoration, the corporations thought long and hard, sometimes engaging in bitter debate over the value of land and traditional ties to it.

The Price of Beauty

Oil spill trustees quickly learned that no public agency had quite confronted the problem before, at least not on Alaska's scale: How to determine the value of 50,000 acres in the middle of nowhere, whose prime selling points weren't favorable zoning or proximity to a large workforce. Instead, efforts to restore the spill zone put a premium on remote, unsettled, often mountainous lands covered not with asphalt but with bears, salmon streams and eagle nests.

Land appraisals typically are guided by recent sales of similar properties and "highest and best use," a standard that may take into account a land's value for logging, mining or housing development. For land whose best asset and greatest value lay in preserving it, no appraisal standard existed.

When government reviewers evaluated parcels that the trustee council wanted to buy, appraisals came in low — a lot lower than Native corporation landowners were willing to accept. For example, government appraisers said 152,000 acres on Kodiak Island owned by Old Harbor Corp. and Akhiok-Kaguyak Inc. was worth $26.3 million, or $175 an acre. The council eventually paid $60.5 million, or $398 an acre.

After months of debate, appraisals were set aside one by one and after closed-door negotiations, prices generally fell between asking and appraisal figures. Money either was passed to Native shareholders as dividends or set aside for shareholders in permanent trust funds. In many cases, easements were adopted to allow traditional uses such as Native hunting.

Prices were unprecedented — but so were the lands' size and splendor.

In the spring of 1998, the trustee council paid $70 million for nearly 42,000 acres on Afognak Island graced by virgin spruce forest. Trustees concluded that Afognak's old-growth forest was exactly what the

LEFT: *What price splendor? Alaska's oil spill trustees paid $70 million for old-growth forest on Afognak Island, northeast of Kodiak Island, declaring it high-value habitat. (Daniel Zatz)*

FACING PAGE: *Critics say too much has been paid for lands that might be better used to develop Alaska's economy. An Afognak Island waterfall graces virgin forest. (Roy Corral)*

Map-making Scientists Guide Oil Spill Cleanup

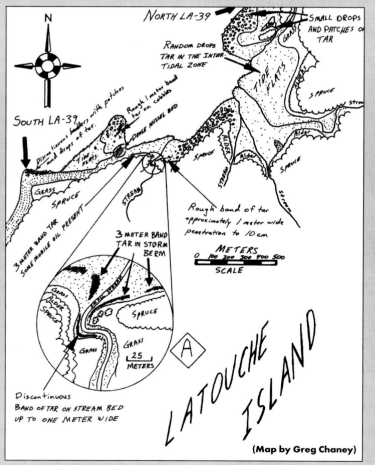

(Map by Greg Chaney)

By Greg Chaney

At first, efforts to clean oil from beaches were inefficient partly because crews lacked accurate maps of oiled beaches. Reality was far more complex than the expanding ink blot of oil shown in the media; oil floated on the sea in large swirls resembling octopuses with trailing tentacles.

To make things more complicated, the shoreline featured steep glacier-carved bays and headlands with pocket beaches strung like pearls between them. Where ribbons of oil touched the coast, a wild variety of concentrations resulted. Oil distribution in some places appeared to be the work of a crazed giant who had run along beaches with a broad paintbrush dipped in crude oil, splattering droplets here, a broad band there or, farther down the beach, dumping the rest in a gooey continuous sheet.

Accurate maps of stranded oil were needed to direct cleanup crews. First, helicopters flew along shorelines to videotape oiled beaches. Tapes were brought to an emergency cartography lab in Valdez and combined with eyewitness reports. Information was then transferred to preliminary computer maps; but maps made from aerial videos were not always accurate. Regions of black wet rocks might appear to be covered with oil, or oil could be missed if it was buried — impossible to find without digging. Maps were then forwarded to shoreline cleanup advisory teams, known as SCAT, which surveyed oiled beaches on foot.

This is where my job started. Cleanup advisory teams were composed of a geomorphologist to map oil distribution, a biologist to document intertidal organisms so cleanup workers would not harm them, and an archaeologist to ensure archaeological sites were not accidentally damaged. (Geomorphology is a branch of geology that focuses on the origin of landforms and how they are modified over time.) As the team's "oil geomorphologist," I used the aerial video maps as a general guide and drew sketch maps of specific sites.

Hand-drawn sketch maps showed reference landmarks to guide cleanup crews to oil we found. Oil concentration also was classified into categories and entered into computerized shoreline maps. These reports were then used to establish cleanup priorities and recommend treatment.

Ultimately the massive response effort was prioritized, directed and carried out using maps as a primary guidance tool. Without rapid deployment of multiple mapping teams, the army of cleanup workers and equipment that descended on Alaska's southcentral coast in 1989 would have been blind and directionless. ●

Geomorphologist Greg Chaney was a consultant in 1989 to the shoreline cleanup advisory team that mapped oil on beaches from the spill's early days through 1992. Chaney, who studied Prince William Sound shoreline processes while earning a masters degree in 1987, works today as a Juneau city planner. His essay is reprinted from Alaska in Maps: A Thematic Atlas *(1998).*

Commercial fishing is a way of life in Prince William Sound. Lands purchased for conservation include top-notch salmon streams to help restore oil spill setbacks incurred by the Sound's fishing fleet. (Kevin Hartwell)

marbled murrelet, whose numbers remain on the decline, needed for nesting. As much as 7 percent of the sea bird's population in the spill zone was lost to the spill; 10 years later, scientists say factors other than the tanker wreck may be slowing the murrelet's rebound.

Native landowners regarded the same Afognak forest and envisioned timber harvests. The trustees' purchase took four years to negotiate, prompting the council's McCammon to smile over Afognak's "gold-plated, diamond-studded" trees.

In Washington, Murkowski took note as well and urged the General Accounting Office to examine the council's spending. In addition to concern that too much was being paid, Murkowski also questioned the purchase of lands never touched by oil. In a report issued in 1998, auditors found little to criticize. While noting that trustees had paid 56 percent over government's appraisals for much of the 360,000 acres purchased to that point, the decisions were not cited as a problem. Murkowski, who said he continued to hear from Alaskans unhappy with the council, said the GAO report stopped short of blessing the trustees' work.

Prince William Sound

In the settlement's early days, the trustee council received hundreds of proposals from scientists nationwide hoping to study oil's

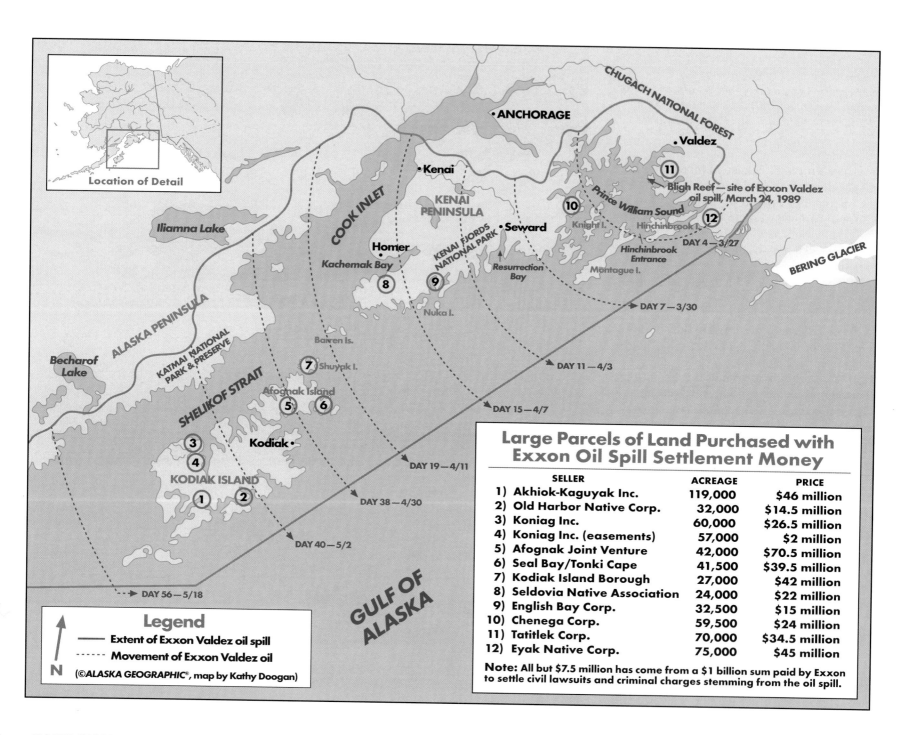

Large Parcels of Land Purchased with Exxon Oil Spill Settlement Money

SELLER	ACREAGE	PRICE
1) Akhiok-Kaguyak Inc.	119,000	$46 million
2) Old Harbor Native Corp.	32,000	$14.5 million
3) Koniag Inc.	60,000	$26.5 million
4) Koniag Inc. (easements)	57,000	$2 million
5) Afognak Joint Venture	42,000	$70.5 million
6) Seal Bay/Tonki Cape	41,500	$39.5 million
7) Kodiak Island Borough	27,000	$42 million
8) Seldovia Native Association	24,000	$22 million
9) English Bay Corp.	32,500	$15 million
10) Chenega Corp.	59,500	$24 million
11) Tatitlek Corp.	70,000	$34.5 million
12) Eyak Native Corp.	75,000	$45 million

Note: All but $7.5 million has come from a $1 billion sum paid by Exxon to settle civil lawsuits and criminal charges stemming from the oil spill.

Legend

— Extent of Exxon Valdez oil spill

- - - Movement of Exxon Valdez oil

N

(©ALASKA GEOGRAPHIC®, map by Kathy Doogan)

effect on everything from killer whales, at the top of the food chain, to copepods, the tiny crustaceans near the bottom. Science that fell within the council's mission of restoration, monitoring and research was considered; the trustees went on to fund nearly $60 million in studies in the Sound. Scientists eventually favored examining the ecosystem as a whole and in 1994, the council began funneling money toward three major ecosystem studies.

Community projects in the spill zone account for roughly $18 million in trustee spending. The Sound's hiking trails have been expanded and salmon stream banks trampled by Kenai River anglers have been restored. A project at Port Dick Creek, on the Kenai Peninsula's southern coast, is aimed at returning the once-productive chum and pink salmon stream to health after Alaska's 1964 earthquake dumped debris and impaired the creek.

Archaeological sites in the spill region have been identified and now are monitored. And a historic Russian Orthodox church site in the abandoned Native village of Kiniklik, in the Sound's northwest corner, may be purchased for preservation.

In Prince William Sound, trustees have spent $115 million for conservation easements or title to roughly 205,000 acres, nearly 600 miles of shoreline and about 175 salmon streams.

Sockeye salmon flooded the renowned Kenai River following commercial fishery closings caused by the spill in 1989. Some $16 million in settlement money has gone to purchase parcels along the Kenai River, where erosion could threaten salmon runs. (Alissa Crandall)

Near Bligh Reef, the council paid Tatitlek Corp., a local Native company, about $34.5 million for easements on some 35,000 acres and title to another roughly 35,000 acres. Much of the land already had been logged but council members argued that acreage was worth preserving for regrowth.

Farther east, near the fishing town of Cordova, trustees offered Eyak Corp. roughly $45 million for a mix of title and conservation easements to 75,425 acres in bays visible from town. No oil made it to the area in 1989, but many bird populations harmed by the spill use the area for nesting, feeding and wintering. The purchase resonated with Cordova-based salmon and herring fishermen who stood by a decade ago as floating oil canceled fishing seasons. Most of the Eyak land was added to Chugach National Forest, which rings much of the Sound.

In the southern stretches of Prince William Sound, the council paid Chenega Corp. $34 million for conservation easements and title to 59,520 acres, including Jackpot Bay. To the north of Jackpot, about 16,000 acres taking in Eshamy, Granite and Paddy bays fell to

Logging was planned on portions of Afognak Island until land purchases by the trustee council made development off limits. Big Waterfall Bay, on Afognak's north central coast, is glimpsed through spruce forest. (Roy Corral)

the state, which has plans for a marine park. A third of the land was added to Chugach National Forest, and part of that is scheduled to become Jackpot Bay Marine Park. The rest was held as conservation easements. Those tracts continue to be owned by Chenega Corp., a Native village corporation, but are jointly managed with the Forest Service. Chenega lands were desirable, McCammon noted, because they include the spill's "ground zero," where shores were thickly oiled.

Along with Chenega and Knight islands, the bays are in the center of Prince William Sound, a stretch that also takes in some of the Sound's most valued salmon streams. Though no immediate plans existed, many observers feared the land eventually would be logged and possibly jeopardize streams.

Chenega Corp. shareholders debated hard before deciding to sell their land. While Chenega did not want to log, it feared the corporation would vanish if the land was left undeveloped. Faced with seeing the trees cut or the land protected, "it made a lot of sense to our shareholders to sell," said Chuck Totemoff, president of the Native corporation. Selling the land to the trustee council in 1997 gave Chenega startup money to pursue tourism, a top source of Alaska jobs.

Kenai Peninsula

Despite pressure to make its first large land purchase within Prince William Sound, the council instead found itself focusing on the Kenai Peninsula, where the state had been negotiating since the early 1970s to purchase nearly 24,000 privately held acres within Kachemak Bay State Park. Acreage was owned by Seldovia Native Association but decades of talks had failed to produce a deal. In the late 1980s, park lovers panicked when the Native association sold timber rights to its land. The tract was visible across the bay in Homer, the end-of-the-road town known for environmental activism. State efforts to buy the parcel intensified and buy-out bills were introduced in the state Legislature, where they stalled.

With oil spill settlement funds in hand, a deal finally was struck in 1992 when trustees set aside $7.5 million toward the overall price of $22 million. The breakdown called for the council to pay a share if the state would spend $7 million out of its criminal settlement funds and another $7.5 million from settlement money paid by Alyeska Pipeline Service Co., operators of the trans-Alaska pipeline.

The council insisted the deal be completed by year's end, that the full price not exceed $22 million and that all timber and mineral rights were included. The

Within seven years of the Prince William Sound spill, area bald eagles numbers rebounded and the eagle was declared "recovered." Authorities say bald eagles will continue benefiting from protected lands that take in 1,200 miles of shoreline and hundreds of salmon streams. (Patrick J. Endres)

state came through; after 20 years of negotiations, title to the land changed hands. Homer was happy.

Another $16 million has been spent on buying smaller parcels on the Kenai Peninsula, specifically lands along the Kenai River. In 1989, commercial fishermen who catch Kenai River red salmon found themselves sitting out when state officials closed seasons, a successful move to keep tainted fish from reaching market. Without fishermen to target the salmon, too many fish reached the Kenai River, a world-class sportfish stream, where their spawn hatched the next year. When too many salmon wound up chasing too little food, experts said the Kenai's salmon population appeared in peril. Meanwhile the Kenai

continued to draw anglers in record numbers. Officials began fretting over each bootprint that eroded the riverbank where salmon fry seek food and shelter.

To try restoring Kenai River health, the council acquired many small parcels along the riverbank and adjacent to it. "We rebuilt damaged riverbanks that had been trampled into hog wallows," says Chris Degernes, area superintendent for state parks. Much of the trustee-purchased land will be left in its natural state while a couple of parcels will provide additional public access to the 105-mile-long river.

Trustee spending on the Peninsula includes community projects, totaling about $51 million, for two new public use cabins in Resurrection Bay outside of Seward and five new cabins in Kachemak Bay. Other projects included hiking trails and campgrounds and a new public dock in Halibut Cove.

In 1997, the trustee council completed a deal with English Bay Corp. for land in Kenai Fjords National Park and the Alaska Maritime National Wildlife Refuge, on the coast near Seward. The $15 million agreement added 30,200 acres within the park and 2,270 acres in the nearby maritime refuge. Government appraisers valued the tracts at about $4 million, an estimate rejected by English Bay.

While corporation members retained hunting rights on some lands, the deal also called on English Bay to use about $500,000 from the sale to fund archaeological and cultural protection work on lands transferred to the park. The purchase included several coves sought out by kayakers and other boaters.

Kodiak Island Archipelago

More than 376,000 acres have been bought or protected by easement on Kodiak, Afognak and Shuyak islands, accounting for the bulk of money spent by the trustees on land. The trustee council had rated much of that land, especially on the north end of Kodiak Island, as "very valuable" wild habitat where salmon spawn and sea birds hurt by the spill go to nest. Another plus: Native corporations that owned the parcels were eager to sell. Other stretches were within Kodiak National Wildlife Refuge and long coveted by environmental groups.

In 1995, two years after buying Seal Bay lands from Akhiok-Kaguyak Inc., the council dealt with the company a second time, buying 73,000 acres in land and conservation easements and another 42,000 acres for $46 million. Tracts are in the heart of

Recreation and tourism were among land uses classified as damaged after the spill; some of the lands purchased with Alaska's settlement money are aimed at enhancing backcountry experiences. Here a paddler revels in solitude in Kenai National Wildlife Refuge. (Ron Levy)

Kodiak National Wildlife Refuge, home to one of the world's largest concentrations of brown bears. More acquisitions in the refuge were to come that year.

In the refuge's northwest corner, trustees purchased nearly 60,000 acres for $28.5 million from Koniag Corp., a regional Native corporation. Agreement terms call for conservation easements on an additional 55,000 acres along the Karluk and Sturgeon rivers until the year 2001; the council has pledged to seek out ways to permanently protect those lands.

On Kodiak's south side, Old Harbor Native Corp. was paid $14 million for nearly 29,000 acres of coastal lands that are prime brown bear habitat. Although brown bear were not on the trustees' injured species list, environmentalists endorsed the Old Harbor purchase because the corporation donated conservation easements on another 3,000 acres and agreed to preserve nearby 65,000-acre Sitkalidak Island as a private wildlife refuge.

With the sale of its coastal lands, residents of Old Harbor, a Native village of about 300, and the Native corporation they belong to, have slowly turned to eco-tourism as a potential source of economic stability. Old Harbor hopes that "tread-lightly" tourism will provide jobs and corporate dividends while not disrupting traditional Native lifestyles centered around hunting and fishing.

Aside from land purchases, the council has spent $2 million for a waste management plan and facilities on Kodiak Island. Recycling and waste disposal sites now are located in a half-dozen Kodiak communities whose isolation was a stumbling block. The council contributed $1.5 million toward Kodiak's Alutiiq Museum, which opened in 1995 and grew out of Native interest in local archaeology. Trustees identified Alaska archaeology as a resource damaged by the spill after cleanup workers wandered remote shores and occasionally looted.

The Kodiak archipelago also has

benefited from council spending for hiking trails, public latrines and new boat docks. At the islands' north end, the council acquired nearly 27,000 acres on windswept Shuyak Island, where the white-washed remains of a World War II lookout post are home to hundreds of terns. Shuyak is uninhabited except for a private wilderness lodge on the island's south side.

Two state park cabins on Shuyak's west side provide shelter in protected Big Bay, a favorite destination of solitude seekers who arrive by small plane and haul in collapsible kayaks. Shuyak's west side

bays are thick with a variety of eiders, red-breasted mergansers, black oystercatchers, harlequin ducks, Barrow's goldeneyes, three-toed woodpeckers and bald eagles. Nearly all of the island became a state park after trustees turned over lands it purchased from the Kodiak Island Borough; prior to that, only Shuyak's west side was park land.

In 1998, the council made its last — and most expensive — purchase in the Kodiak Island area when it paid $70 million for roughly 42,000 acres of dense old-growth spruce forest on Afognak Island's north end. Recalling its habitat ranking system, the council valued Afognak land for shelter it would provide for brown bear, deer, elk, salmon, murrelets and pigeon guillemots. Land was purchased from Afognak Joint Venture, a business owned by Afognak Native Corp. and Koniag Native regional corporation.

Of the total parcel, roughly 6,000 acres were added to Kodiak National Wildlife Refuge; the rest is slated to become part of newly created Afognak State Park.

Looking back on the council's milestones — its multimillion dollar spending for science, new construction and community projects — former council member Deborah Williams is convinced that unspoiled land for people and wildlife tops the list. "This," she said, "is a legacy that will unquestionably be enjoyed." ✈

A staff writer for the Anchorage Daily News, *Natalie Phillips has spent much of the past eight years reporting on oil spill legal battles and efforts to restore Alaska's wildlife and lands. She first glimpsed Exxon Valdez oil while on a western Prince William Sound kayak trip in 1989. Phillips has returned to the Sound each summer since to explore a new stretch.*

FACING PAGE: *Most of the bird carcasses collected after the oil spill were common murres, among species deemed to be recovering in the spill zone by 1999. Common murres perch in summer on a rocky ledge in Kenai Fjords National Park. (Patrick J. Endres)*

RIGHT: *People are part of the restoration equation. Lands purchased with Alaska's $900 million settlement fund are intended, in part, to add back hunting and gathering opportunities marred by the oil spill a decade ago. Here, 3-year-old Mikel Grillo shows off a razor clam at Clam Gulch on Cook Inlet. (Al Grillo)*

View from a Tide Pool: One Writer's Oil Spill

By Charles P. Wohlforth

I was very young the first time I explored an Alaska tide pool but almost no memory is clearer. I can still see the sunshine on Gastineau Channel from Douglas Island, near Juneau: Wizened towers of black rock stand along the shore and sprays of green popweed glisten over puddles and boulders.

Crouching at the crystal window of a tide pool, my eyes squint to distinguish plants from animals and those creatures, like anemone, that seem halfway in between. Hermit crabs in borrowed shells navigate clumsily along the pool's bottom; tiny fish

---

'But even as I struggled in 1989 to convey the spill's enormity, I knew I had failed. We all did. What had been destroyed was too large to express.'

---

dart among the weeds; barnacles and mussels filter the water for invisible food.

As if still pausing there, I smell the salty air, alive with the scent of the intertidal zone that we land dwellers share with sea dwellers just twice a day. That afternoon on Douglas, I learned of life's infinitude. Strange creatures great and small lived every day in the sea and all around us in inconceivable variety and abundance. The natural world was limitless; what a happy and reassuring discovery to hold inside!

I was 25 years old when I landed in a helicopter with other journalists on Prince William Sound's Green Island one sunny spring day in 1989. Although I'd never been here before — few people had — it was a familiar rocky shore, like Douglas Island's. But the smell was wrong: Instead of the fresh, sharp scent of salty rot, here a dense, tarry odor filled the air. Black and brown oil splattered rocks like paint in some places, slathered them inches thick elsewhere like frosting on a cake.

Some tide pools now were oil pools; others were more subtly poisoned so that dying creatures sunk to the floor. Mussels yawned open in the sun, dead flesh stained black. Barnacles slipped from rocks and crumbled at a touch while tiny shrimp-like animals, dead, filled the cracks at the bottom of one pool. Nearly anonymous casualties, I later learned, for these creatures had no common name, only a scientific designation.

Within a half-hour, my colleagues were ready to go. I could hear the helicopter revving. But I felt a responsibility to check under every rock, to witness deaths of unmourned worlds, to stay and never leave. I ran back as the machine bounced on its haunches, anxious to fly away.

Through that summer and into

Journalist Charles Wohlforth exits a military helicopter aboard the USS Juneau, a Navy troop ship used in 1989 to house cleanup workers in the spill's early days. Then-Vice President Dan Quayle was due to inspect a nearby Smith Island beach. (Anchorage Daily News)

the fall, I lived the progression of the Exxon Valdez oil spill — and the futility of cleanup — as a reporter for the *Anchorage Daily News*, the state's largest newspaper. Like many others, I gained financially from the tanker wreck: Journalism launched my career today as a book author and freelance writer for national magazines. But even as I struggled in 1989 to convey the spill's enormity, I knew I had failed. We all did. What had been destroyed was too large to express, the act that caused it too small — a single man's finite mistake, like a boot print crushing a universe in miniature.

The first time I saw the oil spill, it surrounded the tanker like a black wound. Three days later, after the first storm, I flew over the slick, landed on my first oiled shore and returned in little more than an hour. Next day, riding with cabinet officials touring the zone, an hour wasn't long enough for a helicopter to span spilled oil. Later, when I went out with the first bird rescue boat, it took most of a day to traverse the spill. When I visited the Green Island tide pools, we could see the spill's leading edge approaching the exit of Prince William Sound, near Evans Island.

Later that summer, I traveled to the Alaska Peninsula, to see the same spill hundreds of miles to the west. It took all day to fly, first from Valdez to Anchorage, then to Kodiak, then hours by helicopter across Kodiak Island and Shelikof Strait, along mile after mile of uninhabited beach and cliff walls that sped by just below our helicopter's rotor blades. At the end was yet another beach and a little knot of workers, wiping pebbles with rags, one by one.

Again, I had to adjust my understanding of the natural world. I came to see it as a precious liquid in a precariously balanced cup. But to see that, you had to look up close and from far away at the same time.

Look close to see a tiny but complete tide pool world. Catalog every odd little animal within your reach and the larger animals that feed on them. Then pull back to see the whole beach filled with as many pools and pockets of life as there are stars in the sky. Pull back again to see a whole island, a ring of beaches and shorelines curved and folded on each other like lace-fringed paisley. Pull back, higher than an airplane can fly, and see the whole of Prince William Sound, the whole shoreline. Pull back once more to see all of the Gulf of Alaska, all of Alaska itself.

There's the cup, ready to be spilled by a careless hand. ●

An Anchorage writer and ALASKA GEOGRAPHIC® *contributor, Charles P. Wohlforth was a reporter for the* Anchorage Daily News *from 1988 to 1992 and devoted most of 1989 to the Exxon Valdez oil spill, touring the region by boat purchased by the newspaper and piloted by Wohlforth. A two-term member of the Anchorage Assembly, Wohlforth is a lifelong Alaskan.*

Could there be a next time? Industry and government observers agree more safeguards in Prince William Sound as well as heightened monitoring have reduced risk; ongoing work by citizen councils, oil companies and government agencies is aimed at further drawing down risk. (Alissa Crandall)

Out of the Sound: People of the Oil Spill

By Rosanne Pagano, Associate Editor

A map detailing Alaska's known or likely oil prospects is a maze of possible riches, from natural seeps that darken waters near Cape Yakataga 160 miles southeast of Valdez, to offshore rigs that have pumped Cook Inlet for nearly 30 years, to the North Slope where nearly one-fourth of U.S.-produced oil originates. Less well known are potential deposits that dot Alaska from the Copper River region 200 miles east of Anchorage, to Bethel in southwest Alaska, throughout the Interior, and into the Bering Sea where the far-flung Bristol Bay, St. George and Navarin basins are found.

But prospectors soon realize why much of Alaska's wealth remains underground: Unlike the classic gushers of Texas, Alaska

FACING PAGE: *To conserve and protect: The Alaska SeaLife Center on Resurrection Bay in Seward, built partially with oil spill settlement money, combines marine wildlife research with a public aquarium to help tell the oil spill story. An outdoor deck overlooks a sea lion at play. (Al Grillo)*

oil often is hard to tap, trapped in deposits hundreds of miles from lucrative markets. Despite these obstacles, the state's dependence on oil was sealed in 1968 when Atlantic-Richfield announced its discovery at Prudhoe Bay. The North Slope reservoir, the nation's largest, was found to hold as much as 25 billion barrels of oil and an estimated 30 trillion cubic feet of natural gas.

This time, geography would not limit destiny. The state, which leased North Slope tracts to developers and would realize royalties from production, shared industry's determination to get Alaska crude to market. But how?

Fueled by the end of cheap foreign oil and the rise of the environmental movement, national debate soon focused on proposed Alaska-oil shipping routes, the topic of heated public hearings. Suggestions included a shipping route across the Arctic Ocean, relying on icebreakers to reach East Coast markets, and a pipeline that would cross Alaska and Canada to deliver oil to the midwestern United States. Steadfastly opposed by environmentalists was yet another option — a trans-Alaska pipeline from the North Slope to the ice-free port of Valdez, where oil

would be shipped south by tanker.

Alaskans eager for a construction boom favored the all-Alaska route; so did federal reviewers who concluded that new, tougher standards would reduce the chance of mishaps as tankers navigated Prince William Sound's storms and icebergs.

Alaska Natives, owners of much of the land the 800-mile pipeline would cross, resolved their objections to the project in 1971 when Congress passed the Alaska Native Claims Settlement Act, granting $962 million to Native groups and title to 44 million acres. Two years later, Congress again removed a hurdle by exempting pipeline construction from many requirements of the National Environmental Protection Act of 1969.

Among those tracking environmental concerns was a displaced Californian who, in 1968, believed Alaska could do better and spoke up at Anchorage hearings. "I lived through all that," recalls Mike O'Meara, a curator for the past 10 years at Homer's Pratt Museum. "The whole debate included discussion of the dangers of the marine route. There were good suggestions back then that ultimately were ignored, to our peril."

an updated version of "Darkened Waters" continues to attract visitors curious about the spill's ripple effects in Alaska. For the anniversary year, displays will be on view in Seward at a University of Alaska marine center and at the Pratt.

For O'Meara, who says his work on "Darkened Waters" began as a six-week temporary stint and developed into a career, the issue of safely moving Alaska oil remains an abiding interest 31 years after he first moved north. Upgrades in navigation, tanker operation and escort ships mandated since the accident are a significant legacy, he says: "We all breathe easier knowing they're there."

Oil Spilled, Archaeology Found

Efforts to protect Alaska's oil spill zone, where much of the state's population lives and tourism flourishes, have not focused only on conservation for the future. Along with millions of dollars spent on commercial fishery research, wildlife study and land purchases, Alaska has used some $7 million in settlement money to document and preserve its past.

Like the state's untapped basins of oil, archaeological treasures abound in much of Alaska where ice, soil composition, relatively stable year-round temperatures and virgin land combine to produce favorable conditions for preserving relics. Alaska's spill zone, which takes in the Kenai Peninsula's outer coast and Kodiak Island, has long been regarded as a rich venue for

O'Meara says he was a natural choice as exhibit curator in 1989 when the Pratt Museum began planning an oil spill display to open within two months of the tanker grounding. "I had historical background," he says. "I knew this story firsthand."

At first, "Darkened Waters" began as an attempt to explain the wreck and aftermath to townspeople in the spill zone, some of whom joined cleanup crews and had witnessed oil's destruction. For all their reliance on oil — Alaska's permanent fund

dividend, paid annually to every resident from the state's oil-wealth savings account, is a popular bonus — few Alaskans had spent much time pondering oil's toxicity before 1989.

"Darkened Waters," with its discussion of Alaska's oil history and images of oil-soaked wildlife, changed that. A traveling version of the exhibit eventually was put together, toured Alaska and was displayed in 15 cities outside the state, including at the Smithsonian Institution. Ten years later,

archaeologists whose only wish has been more funding for field work at remote sites.

Douglas Reger, an archaeologist with the state Department of Natural Resources, has worked since 1991 on spill zone archaeology that he says probably would not have been pursued except for interest in, and funding for, documenting oil's damage. In addition to Prince William Sound, archaeological study has concentrated on Nuka Island, a Gulf of Alaska site roughly 130 miles northwest of Kodiak, and Shuyak Island, a short trip across Shuyak Strait north of Afognak Island near Kodiak.

Reger says findings in Prince William Sound show the sites were inhabited earlier than previously thought. At Nuka Island, artifacts dating to between 1,000 years and 600 years ago were found in an intertidal zone's peat deposit that had helped preserve the material. Grooved adzes used for splitting wood were uncovered as well as a glass trade bead.

Oil spill archaeology "did fill in some gaps in knowledge," Reger says. For instance, researchers found that certain Kodiak Alutiiq cultures existed at about the same time on Shuyak Island as well. "We suspected that, but had not been able to document it," Reger said. Dates on Shuyak artifacts also fall between 1,000 years and 600 years ago. Today much of Shuyak Island is set aside as a nearly 50,000-acre state

"We all breathe easier," notes Pratt Museum curator Mike O'Meara, referring to safety upgrades in place since the 1989 oil spill. In Prince William Sound, these visitors thrill to glaciers and wildlife in College Fiord. (Patrick J. Endres)

park increasingly popular with kayakers.

Some archaeological sites within the spill zone first were uncovered when damage assessment teams in the early 1990s walked the shorelines. Others were stumbled upon by cleanup workers who, authorities say, looted or damaged the sites. Despite those losses, a top concern following the accident was how Exxon Valdez oil might foul efforts to carbon-date wood or bone; studies have since shown that fear to be unfounded if laboratories are alert to pollution and

undertake cleaning. In a report released in 1993, archaeologists noted that damage to spill zone sites "appears to be from erosion or vandalism, rather than from direct oiling."

Once archaeological resources were formally identified as injured by the spill, settlement money could be spent for repair. One estimate put the restoration sum for archaeology at roughly $872,000, including site repairs and 10 years of monitoring oil's effects. For a cost study in 1993, 35 sites in

Prince William Sound and the Gulf of Alaska were reviewed, including one place where damage was deemed "unrestorable."

Field work so far has turned up stone artifacts as well as pieces of wood and bone. On Shuyak Island, seeds were located; a piece of carved wood from a Prince William Sound site may once have been part of a kayak. "We know they had centralized winter camps, and hunting and fishing camps," Reger says of early Gulf of Alaska people, who also were known to return to the same area year to year. Pinpointing how each of the newly uncovered sites was once used remains a puzzle.

Reger's most recent season in the spill zone was summer 1998 but he says that as oil spill money recedes, little additional field work is planned any time soon. Ahead is a final write-up of findings as well as several more years of monitoring vandalized sites, an effort that in the future could rely more on local stewards.

Big Hopes for Littlenecks

When pioneers migrated to the American West, the allure was independence that came from living off the land. In coastal Alaska, independence today is living off a sea so abundant that villagers say the table is set when the tide goes out.

Clams and cockles, mussels, octopus, seaweed, chitons and shrimp are only a handful of the subsistence foods harvested for centuries from waters within the Exxon Valdez spill zone. But in the 12 months following the tanker wreck, researchers noted that harvests of traditional foods in coastal villages dropped by half in some places and by three-fourths in some Kodiak Island communities, prompting the settlement trustees to rank subsistence hunting, fishing and gathering as an oil-damaged resource.

While subsistence living in Alaska dates

Clipped Hair, Spilled Oil

When crude oil flowed from the Exxon Valdez tanker, a flood of entrepreneurs worldwide suggested novel ways to separate oil and water. Cleanup officials had tried burning the slick, corraling it and treating it with chemicals but in the end, no one could hold back the sea; currents and storms dispersed the oil, touching off a $2 billion cleanup that, the Environmental Protection Agency says, managed to retrieve 23 percent of the 11 million gallons lost; state officials put the estimate at less than 10 percent.

At least one entrepreneur says it didn't have to be that way.

"The light went off," recalls Alabama hair stylist Phillip McCrory, who says his oil-absorbing idea came to him after seeing pictures of Alaska's oiled otters.

Backyard tests were promising: After dumping a gallon of used motor oil into his son's wading pool, McCrory noticed the water cleared in a few minutes after he added a filter made of tights stuffed with human hair.

NASA's Marshall Space Flight Center decided to give McCrory's hair filter a try when diesel oil spilled into a water ditch at the Huntsville site. McCrory — whose customers include some flight center employees — fashioned a filter from 16 pounds of human hair and a barrel. The result: Filtered water retained oil at the rate of 17 parts per million, sufficiently clean to be disposed into a sewer.

More laboratory tests were done, and some authorities estimated the Prince William Sound spill could have been soaked up in about a week's time using roughly 1 million pounds of human hair in filtering pillows. McCrory, who says his idea works because hair cuticles give oil in water something to stick to, says that at least 10 tons of hair are disposed of daily at salons nationwide.

In Anchorage, salon owner Sandee McDowell sends clippings of clean hair to McCrory once a week. "He said coming from Alaska, because of the oil spill here, made it special," McDowell said.

Stylist Lisa Alleva cuts hair at Anchorage's Chez Ritz Salon, which saves clippings for an Alabama man who devised an oil-collecting filter made of human hair. (Anchorage Daily News)

ABOVE: *A Seward hatchery cultivates littleneck clam larvae to eventually seed village beaches within the spill zone. While there've been no harvests yet, seeded clams appear to be growing faster than wild clams. (Exxon Valdez Oil Spill Trustees)*

RIGHT: *The Qutekcak hatchery hopes someday to attract the interest of Alaska's mariculture industry. "Oyster Suzie" shows off a Prince William Sound harvest. (Patrick J. Endres)*

to the earliest peoples, little formal research had been done on the purity of foods consumed. There was hardly a need: Sparse population combined with scant industrial pollution tends to keep much of coastal Alaska nearly contaminant free, the same conditions Western settlers encountered more than a century ago but soon obliterated.

In Chenega Bay, a western Prince William Sound village in the spill's direct path, the change from pristine to ruin occurred literally overnight. For some, the loss was reminiscent of another Good Friday, in 1964, when southcentral Alaska

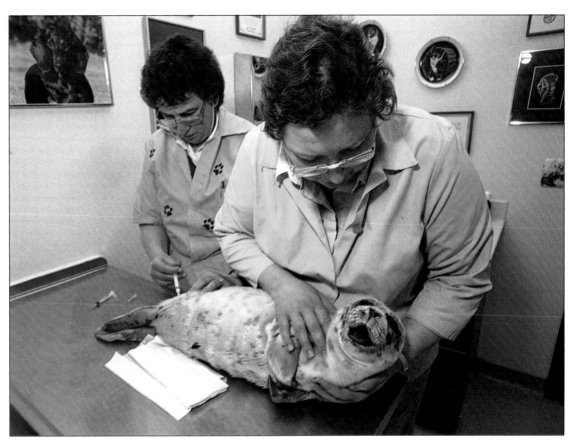

was rocked by a "moment" magnitude 9.2 earthquake and Chenega lost nearly a third of its 68 residents in a tidal wave. Villagers, many of them Alaska Natives, dispersed after the destruction, returning to a site south of the original settlement just five years before the tanker went aground.

"When spilled oil arrived on their shores, (the villagers') reacquaintance with their fishing and hunting practices was disrupted," writes Nancy Lord in the book *Darkened Waters* (1992), to accompany the Pratt Museum's oil spill exhibit. "A strong sense of loss again washed through the community." Chenega, 60 miles west of the grounding, and the village of Tatilek, about five miles northeast, were among towns reporting declines of nearly 60 percent in traditional food harvests following the spill. Barges of free food sent by Exxon were appreciated, Lord writes, but the unfamiliar meats and processed foods failed to satisfy: "Being given foods — even fresh fish and clean seaweed — was simply not the same

as living a life that involves every step of providing what goes on the table."

Enter the planted littleneck.

When both subsistence Alaskans and weekend diggers began avoiding spill zone clams, hatchery workers in Seward set to work to offset the loss by raising littleneck clams and seeding beaches. "It's a lot of hands-on work," says Qutekcak Hatchery biologist Jon Agosti, who monitors hatchery-reared juvenile clams known as "spat." "It's easy for pathogens to wipe out a group."

The hatchery, which moved last year into a $3 million, state-owned mariculture technical center in Seward, is run by the local Qutekcak° tribe. Funding for the center came from Alaska's oil spill settlement.

Qutekcak began work in 1995 to start a spill zone littleneck clam harvest where one hadn't existed before, to augment the subsistence take. "Planted" clams placed in beaches in 1996, the project's first year, were found a year later to be growing faster than wild clams. Results so far are promising, but Agosti says he wants a few more years of good growth before declaring the effort a success. "Then we could say with confidence that it's a three-to-four-year

crop, versus our worst fears in the beginning that it might be a six- or eight-year crop." Slow-growing clams would do little to fill out a subsistence larder.

Tiny clam larvae are fed three types of plankton and housed in six, 30,000-gallon tanks cleaned daily. Larvae are trapped on screens where their health and abundance are noted before being placed into a fresh tank. So far there has been monitoring of the planted clams but no harvests.

Agosti says commercial shellfish growers, pursuing a promising but relatively new Alaska industry, also could see longterm benefits from the Qutekcak project. Many of Alaska's commercial shellfish farmers today raise oysters; Agosti says the new mariculture center is prepared for bivalves other than clams. "We want to produce large numbers of spat for sale to growers," he says. "This facility was built with an eye for the future, when the industry will be substantially larger and they'll grow additional species."

Prevention First

After a spike in 1991, when roughly 325 million gallons were spilled worldwide, both the number of oil spills and gallons lost have dropped substantially. The Massachusetts-based research group Cutter Information Corp. reports the average amount of oil spilled globally now stands at 98 million gallons a year, while the average number of spills a year is 221. In 1997, 136 spills — both inland and marine — were reported, the smallest number of incidents since 1968. Cutter also notes the amount of oil lost in 1997, roughly 49 million gallons, was among the lowest spill volumes in 30 years, and there were no spills of 10 million gallons or more.

"Policymakers and the oil transport industry would like to speculate that the apparent downward trend (is) indicative of the success of spill prevention efforts," notes Cutter's *Oil Spill Intelligence Report* (1997). While acknowledging the role prevention efforts play, the report also concludes that more years of data are necessary to outline a trend.

But it doesn't take a tanker or barge wreck for oil to reach the water: The world's largest spill, estimated at 240 million gallons, was an act of environmental terrorism when the Iraqi army set fire to

500 oil wells in Kuwait during the Gulf War in 1991. A far more common pollution source — accounting for more than one-third of oil entering the water — is a combination of runoff and municipal and industrial waste. Tanker operations account for about one-fourth of oil in the water, while tanker accidents rival natural pollution sources: Each accounts for about one-tenth of oil entering the waters each year.

In Alaska, remote towns touched by the 1989 oil spill quickly learned that preventing marine pollution warranted at least as much attention as cleanup. In 1996, the Sound Waste Management Plan was adopted by several spill zone communities seeking to resolve common problems such as overstuffed landfills or disposal of hazardous waste and scrap metals. In 1999, Alaska's oil spill trustees set aside nearly $2 million for the waste management project. Old Harbor villager Jeff Peterson, who lives on Kodiak's remote southeast shore, helped organize a local project after noticing that old boilers and furnaces from a local housing complex had been hauled to the village landfill and left to rust.

"That's 200 bulky things up at the dump," he told an interviewer. "Why can't we just put them on the barge that brought the new ones?" Annoyance gave way to grassroots organizing and by spring 1998, seven

Not everyone can scout for sea stars, seen here in their natural setting on the Kenai Peninsula, but Alaska's wildlife is a big draw for visitors. To satisfy curiosity fueled partly by the oil spill and partly by nature-based television programs, the SeaLife Center brings wild Alaska inside. (Patrick J. Endres)

outlying Kodiak Island communities had joined the Kodiak Waste Management Plan. Borough authorities hope the alliance ultimately will lead to safer waste disposal — an expensive, never-ending problem for towns off Alaska's road system.

Among the most widespread pollution hazards in villages is waste oil from commonly used machines such as generators, snowmobiles and boats. Kodiak Island Borough managers concerned about safe handling and disposal will soon try a

new waste oil incinerator that can be loaded on to a boat to loan to villages.

Sound Work:
Students Learn from the Spill

Roger Sampson was not about to let an oil spill go to waste. As superintendent of Chugach School District, covering 22,000 square miles from Whittier on Prince William Sound to Yakutat on the Gulf of Alaska coast, Sampson took notice when world renowned scientists targeted the spill

FACING PAGE: *Ripple effects: Cordova is home port to the Sound's commercial salmon fleet but fishing isn't the economic engine it once was. In a 1999 report by the state Labor Department, economists said Cordova's post-spill economy has stagnated since 1989. An estimated one-third of the town's workforce is tied to fish harvesting or processing, sectors that have weakened since 1991 following harvest and price declines. (Patrick J. Endres)*

zone and turned it into a living laboratory. "The traditional approach would have been to invite these scientists into the classroom as a guest speaker," Sampson says. "That's good for about 35 minutes."

Instead Sampson and other Chugach administrators favored an alternate approach, one that would treat students whose towns and families had witnessed oil pollution as scientists-in-training. "We believed all along that learning takes place beyond the walls of the classroom," he says. "The oil spill gave us a chance to walk the talk." In 1995, Youth Area Watch began.

Aimed at students from grades 7 to 12, the program pairs oil-spill scientists with students selected in part for their willingness to share what they've learned with classmates. Science-loving students aren't the only ones gravitating to the project. "Hands-on learners love it," Sampson said. "We find they become avid readers because now there's a reason to apply what they've learned, do the research and communicate." The district counts 180 students in all, including 50

children at its largest school in Whittier.

With a grant from the oil spill trustees, Youth Area Watch sends students into the field to work with researchers on tasks ranging from water sample gathering to groundfish counts to weighing juvenile herring, measuring about 100 fish to the ounce. Students also have a chance to travel to a federal research lab at Auke Bay near Juneau, where analysis is done. One Chugach graduate who took part in Youth Area Watch became expert in gathering biosamples from harbor seals, among species whose numbers have yet to rebound following the oil spill. Sampson says that student has gone on to pursue a biology degree at the University of Alaska Fairbanks.

Unplanned benefits surfaced as well. By exposing school-age children to rigors of the scientific method, the village balance of

Home at Last: Museum Captures Kodiak's Past

It didn't take an oil spill to nudge Kodiak into learning more about the island's Alutiiq people who inhabited the region 7,500 years ago, living in sod houses and fashioning some of the earliest slate tools.

It did take a spill in 1989 — and settlement money that flowed from it — for Kodiak to realize a longstanding dream of housing its bountiful archaeological artifacts at home. With a $1.5 million grant from the Exxon Valdez trustees, an Anchorage-based council overseeing the $1 billion settlement fund, Kodiak Native groups helped develop a 5,000-square-foot Alutiiq Museum housing some 100,000 objects. In addition to public exhibit areas opened in 1995, the museum also features professional storage vaults and a laboratory. Many artifacts were excavated before the oil spill.

The museum is one of just three buildings constructed with settlement funds (the others are the SeaLife Center and a mariculture hatchery, both in Seward) Today, the Alutiiq Museum prides itself on research as well as fostering community-based archaeology that already was under way in Kodiak before museum construction began in 1994.

"Archaeologists have a bad reputation for taking things away from communities," notes Amy Steffian, Alutiiq Museum deputy director and an archaeologist herself. "We needed a place to bring these collections back home."

With its relatively temperate climate and easy access to fishing, Kodiak Island has for centuries been a crossroads of Alutiiq culture that stretched west from Prince William Sound, to the Kenai Peninsula, to Kodiak Island and the Alaska Peninsula. The groups — known as Chugach, Unegkurmiut and Koniag, respectively — spoke slightly varying dialects of the Alutiiq language; they shared a pattern of living in large coastal villages and hunting marine mammals from skin-covered kayaks. And as Steffian notes, they left behind an abundance of artifacts .

"We have almost 850 known archaeological sites" in the Kodiak region, Steffian said, adding that Kodiak accounts for about 4 percent of all known sites in Alaska. "Considering that the area makes up less than 1 percent

FAR LEFT: *The Alutiiq Museum on Kodiak Island is a repository for regional artifacts once housed as far away as the Smithsonian Institution. Opened in 1995, the museum helps kindle interest in Alaska Native heritage. Construction funds came in part from Alaska's oil spill settlement. (Amy Steffian)*

LEFT: *Museum-organized digs attract local volunteers, including students who earn high school or college credit. Volunteer LaRita Laktonen helps excavate the Blisky Site, on Near Island within minutes of downtown Kodiak, in 1997. (Amy Steffian)*

The museum's collection, which includes some 100,000 objects, features rare pieces of prehistoric carvings unearthed within the Kodiak archipelago. An ivory bird carving (right) and an incised pebble (far right) date from 700 to 400 years ago, both from the early Koniag cultural period. Museum curators say the bird carving probably was a hunting visor decoration while the pebble depicts a figure dressed in ceremonial regalia. (Both: Patrick Saltonstall / Afognak Native Corp. Collection)

of Alaska's total land mass, that's a very high percentage."

Formal archaeological digs on Kodiak date at least to the 1930s when anthropologist Ales Hrdlicka excavated several hundred burials and, despite Native objections, removed thousands of artifacts for the Smithsonian Institution. (In 2000, the museum will display the Smithsonian exhibit "Looking Both Ways," which includes Alutiiq material collected in the late 1800s including clothing, weaponry and masks.)

Steffian says the sheer abundance of Alutiiq artifacts still remaining on Kodiak, combined with heightened pride in Native culture, led to action by the settlement fund trustees who had identified spill zone archaeology as a resource damaged by the tanker accident and in need of restoration.

"Before the spill, there was no place to store Alutiiq artifacts,"

Steffian says. "They were scattered through museums and private collections." Then in 1991, after years of pressure, centuries-old funerary objects and human skeletons stored at the Smithsonian were repatriated at Kodiak; other Alutiiq artifacts have been unearthed in Kodiak garages. "Because of the spill," Steffian said, "these artifacts without a proper place to live have a home."

Steffian, who says hers is a "dream job," serves as the museum's lead archaeologist, teaches archaeology at a Kodiak college and helps organize museum-sponsored digs that rely on local volunteers. Recent field trips — including students who earn high school or college credit — have taken crews to a site on Near Island, one-half mile south of the city of Kodiak, where hunting gear and tools for processing skins were uncovered. In 1998, volunteer archaeologists traveled just 14 miles from downtown Kodiak to Zaimka Mound, where ancient sinkers used to weight fish nets in the water were found. Both sites

are estimated at between 3,000 years and 5,500 years old.

Zaimka Mound on Native-owned land rose in priority after joyriders in all-terrain vehicles defaced the site, kicking up artifacts in their dust. To methodically locate relics in place, volunteers used hand tools to dig a hole 5-feet deep and 19-feet square at Zaimka ("meadow" in Russian); materials

found there now are stored at the museum. Artifacts include arrowheads and scrapers, giant stone mauls for stake-pounding, and remnants of sod houses dug below the surface of the ground.

"People want to get dirty and dig stuff up," Steffian says of her volunteer archaeologists. "Every spring we have an exhibit of materials we've found, then we take off for another season." •

power has shifted a bit. About half of the school district's students are Alaska Native, a culture guided by the knowledge of local elders. Sampson says that since the oil spill and Youth Area Watch, Chugach students are becoming expert on the condition of traditional food such as mussels, shrimp and roe. That knowledge, Sampson says, allows students to more fully take part in community discussion about food resources.

The Chugach district is planning now for the inevitable dwindling of oil spill grant money by helping support Youth Area Watch with its own funds and other grants. Science research also will ebb as settlement money comes to an end but Sampson believes there will continue to be a need for baseline studies — some of them, he hopes, led by Chugach students who got their start with Youth Area Watch.

"They understand," Sampson says, "that they not only have some control but they have some responsibility as to how to manage their own resources, as part of their culture and their every day lives. They know they don't have to rely on someone else to provide expertise."

At the SeaLife Center, Science Tells the Story

Where the two-story, $55 million SeaLife Center sits today, volunteers gathered in 1989 to clean oily sea otters. Ten years later the animals' numbers remain depressed in the spill zone's heavily oiled areas and today, the telegenic sea otter has become a symbol of the tanker accident that many Alaskans vowed never to forget.

The Alaska SeaLife Center houses that collective memory. A unique cold water research laboratory and public aquarium, the 115,000-square-foot site in downtown Seward tells the oil spill's story through the animals it displays and science it supports.

"The oil spill focused our attention on Seward" as a site, said Kim Sundberg, SeaLife Center executive director. "But

LEFT: *Fresh-caught herring for oil spill research are stored in a bucket. (Roy Corral)*

FACING PAGE: *A Prince William Sound fisherman sizes up a rockfish caught by a tourist. Rockfish, which are harvested commercially, are the target of a genetic study begun in 1999 and funded by the oil spill trustees. The fish is listed as an injured species based on carcasses retrieved, but extent of damage is unknown. (Patrick J. Endres)*

LEFT: *Some Alaskans couldn't resist turning the oil spill saga into T-shirt slogans. A vendor displays some Valdez-made shirts in 1989. (Cary Anderson)*

ABOVE: *Archaeology was included on an inventory of Alaska resources injured by the spill after authorities said cleanup crews had disturbed some sites. Restoration funds underwrote field work in 1994 within Chugach National Forest, which borders the western Sound. (Exxon Valdez Oil Spill Trustees)*

the genesis really goes back 20 years with a grassroots effort to get more marine research facilities in Seward."

Eighty miles south of Anchorage on Resurrection Bay, Seward's coast was untouched by Exxon Valdez oil. The port town was chosen as an animal rehabilitation site because it lies about halfway between Prince William Sound to the east and the spill zone's western half; animals from either zone could be delivered there. "Sea otter rehabilitation was a short-term response to a crisis," says Sundberg, a retired state biologist who worked on oil spill damage-assessment studies. "In the

long-term, what was needed was a marine research lab."

State officials and oil spill settlement trustees agreed. In 1994, the trustees awarded $25 million to the center for its research and animal rehabilitation units. Funding also came from the city of Seward and private donors. On opening day, May 2, 1998, hundreds of people swelled Seward's Fourth Avenue, where shopkeepers welcomed the tourism spillover. "I saw things I never dreamed of," one visitor wrote on a SeaLife Center comment card. Another said, "We had a great time while gaining knowledge." In addition to an exhibit

chronicling events in the decade since the spill, the center's displays are full of crabs, puffins, sea lions, seals and "touchable" marine life such as sea stars. "This is really cool!" one visitor wrote.

Seward was an attractive setting not just because of its oil spill roots but because of its access to unspoiled Resurrection Bay, whose waters circulate through SeaLife Center labs and tanks at the rate of 5,000 gallons a minute. Planners had considered Anchorage, the state's largest city on Cook Inlet, or Cordova on Prince William Sound but sea water at both sites turned out be too high in sediment.

Oil spill studies at the center include a recently completed river otter project, aimed at learning the effects of ingested oil. Preliminary results revealed that low levels of contamination may have "profound effects" on otters' ability to dive and could reduce their ability to capture prey. Fifteen male otters were trapped for the SeaLife Center study and divided into three groups. One group was fed a usual diet while food given to the other two groups contained small amounts of oil. Anemia and elevated levels of liver enzymes were detected in the oiled otters, scientists said. After being returned to a usual diet, all 15 animals were released to the wild in March 1999.

Other oil spill science at the center focuses on establishing a colony of pigeon guillemots; health testing on captive harbor seals; and genetic testing of pink salmon. Trustees say all three species have yet to recover from oil spill losses

"The oil spill taught us first of all that we had very inadequate knowledge of Alaska's marine wildlife," Sundberg said, noting the lack of baseline population numbers for many species. "Oil, fishing, ocean temperature, increased tourism and boat use — there are a number of factors

Oil marred the harbor seals' key haul-out areas in Prince William Sound but a population decline that continues today began before the Exxon spill. Harbor seal studies at the SeaLife Center focus on blood chemistry and diet, among other survival factors. (Patrick J. Endres)

that affect fish and wildlife. Scientists just don't really know what's causing changes in these populations."

The center also is conducting extensive study into the western population of Steller sea lions, an endangered species which has seen a 90 percent decline in the past 30 years. (The western population group extends from Prince William Sound to the Aleutian Islands.) Three Steller sea lions live at the center, where research has focused in part on understanding the

animals' diet. For instance, scientists know that herring was a key component of the Steller sea lion diet 30 years ago; today, the target fish is pollock. By offering varying diets including pollock, herring, salmon, Pacific cod, squid, capelin and other fish, SeaLife Center researchers hope to learn if food availability has played a role in the population decline.

The two-year project, which is not linked to the oil spill, is ambitious: Adult Steller sea lions can weigh 1,000 pounds or more

and have their own elevator and extra-wide corridor at the center to move between indoor and outdoor lab areas. A closed-circuit video system permits underwater taping. Study results could be applied to Alaska's $1 billion commercial pollock fishery, which is operating under federal restrictions to protect Steller sea lions. In a 1998 report, the National Marine Fisheries Service noted that studies on captive animals "will likely form a basis from which dietary requirements of wild animals can be determined and understood."

SeaLife Center admission money is applied to research, to underscore the center's twin goals of serving both the public and science. In 1998, roughly 4,500 students — including some overnight groups that rolled out sleeping bags adjacent to the floor-to-ceiling exhibit tanks — visited the center. Sundberg says the center's emphasis on displaying and explaining its research builds a constituency: "They'll support the research to conserve animals that are important to all of us."

Cormorants perch on a small rocky island in southern Prince William Sound. (Patrick J. Endres)

ABOVE: *Spill response workers read a news account of the out-of-court settlement between Exxon and state and federal governments in 1991. The oil company paid a $150 million criminal fine, the largest ever imposed for an environmental crime, and $900 million to settle civil litigation and establish a restoration fund. That sum is being paid in installments over 10 years ending in 2001. (Patrick J. Endres)*

RIGHT: *Calm pervades a sunrise view of Cataract and Surprise glaciers in Harriman Fiord, about 80 miles west of Valdez. Ten years later, the spill continues to resonate in the fishing village of Cordova, subject of ongoing research undertaken by Alabama sociologist Steven Picou. He says reports of depression and broken families are reminiscent of survivors' stories from environmental disasters such as Love Canal or Three Mile Island. (Patrick J. Endres)*

Lessons of an Oil Spill: Alaska and Beyond

By Ernest Piper

After the Exxon Valdez ran aground in 1989, Alaska endured entrepreneurs and experts; legislators and regulators; the press and the politicians; opportunists and activists; scientists, accountants and lawyers; the winners, losers and bewildered — all of whom mostly came and went.

The accident's durable legacy is the federal Oil Pollution Act of 1990, a 91-page effort to codify demands that a massive oil spill never happen again in U.S. waters. Today there are expanded contingency plans and sterner financial liability laws. Alaska's experience has informed planners worldwide, led to the founding of a multimillion-dollar aquarium and marine research site at Seward, and generated volumes of technical literature that account for one-fourth of the holdings

FACING PAGE: *The Valdez marine terminal is loading point for tankers carrying Alaska North Slope crude, a waxy oil extracted on the North Slope and pumped through the 800-mile trans-Alaska pipeline. (Kevin Hartwell)*

at an Anchorage public library.

But have things truly changed? A surprise practice drill called in Prince William Sound by state officials in January 1999 revealed that some response crews had not been fully trained, some were tardy in arriving and others were unclear about their duties. Alyeska Pipeline Service Co., which runs the Valdez marine terminal and maintains spill response teams, has acknowledged some shortcomings but declared that overall its crews were ready. The drill was held after Alyeska said low oil prices prompted plans to lay off five of 50 response workers. "We have concerns about their ability to conduct an initial response," the state Department of Environmental Conservation said.

Casualty investigations reviewed in 1996 by the National Research Council, which advises the federal government on science and technology, reveal that most oil spills occur in benign weather, involving well-equipped, seaworthy ships with competent crews. Studies by the National Transportation Safety Board show that most accidents can be traced to lapses by deck officers that either went undetected or not communicated to others until too late.

Things were no different on March 24, 1989 when the Exxon Valdez, traveling about 14 mph and without an escort vessel, swung out of established shipping lanes to avoid icebergs northwest of Bligh Reef. That night, the single-hulled, 987-foot tanker was not the problem; built in 1986, it was among the newest vessels hauling Alaska North Slope crude. The ship's hull was sound, its electronics first-rate. All motors, gears and computers that go into running a tanker were at or near the top of industry standards.

In 1989, the weak link in tanker safety was people. In 1999, the lesson of the Exxon Valdez is that safety hinges on diligent crews, government regulators and elected officials as well as citizens who know the price of complacency. And if oil is spilled in the water? Does improved technology exist to mop 11 million gallons of sludgy crude from 40-degree waters, much of it productive fishing grounds?

Equipment to skim oil from water today is not much different from gear or strategies available in 1989. The technique of burning oil floating on water has advanced, but experts agree its success still depends on weather and sea conditions. Improved

chemical dispersants, which work by breaking a slick into droplets, pose less of a threat to the environment than previous formulations, but more field tests are needed. Again, weather and seas are factors and good results from dispersants, dropped from airplanes, still rely on skillful flying and spotting.

If another accident polluted Alaska shores, cleanup officials trying to sop oil while also limiting harm to the environment might find fewer options at hand than were attempted in 1989. Government and industry studies over the past 10 years have questioned long-term benefits of

intrusive cleanup and now suggest that, in many cases, shorelines simply be left alone. And if there is cleanup, it will still be slow, grubby and expensive.

But while technological stumbling blocks remain, significant advances have been made since 1989 as industry, government and watchdog groups become more skilled in managing risk that comes with moving oil or mopping spills.

Here's what has changed:
• It should be more difficult for tankers to run aground. Enhanced radar and satellite data used by the Valdez Coast Guard permit tankers to be monitored for longer

distances; better communications gear is in place and the Sound's tanker escort system is among the most extensive in the world.
• If oil is spilled, retrieval should be prompt. While there aren't many new techniques, there is more gear on hand, stowed close by.
• A tested incident management plan will be applied to limit chaos touched off by a big tanker spill. The next accident response will find agency and industry authorities working under a clearer chain of command, with better plans and support units and assigned tasks.
• If shorelines are polluted, more oil will be left on the rocks, prompting renewed debate over how to weigh competing interests of science and public policy.

Preventing a Next Time

Accidents are named for the tankers that carried the oil, a bit of blame-shifting that tends to mask an oil spill's cause, obscure responsibility and limit the search for solutions. A 1993 technical paper suggests that 71 percent of groundings can be traced to error by a vessel's bridge crew. (The bridge is the vessel's "cockpit" or control center.) But almost from the earliest hours after Exxon skipper Joseph Hazelwood reported that his ship had fetched up hard aground and was leaking oil, the official response has been to add regulations, equipment and technology, much of it borne of understandable reaction to oiled, dying wildlife.

Then and Now: Progress in the Sound

1989	1999
No ship escorts required except when loaded tankers transited Valdez Narrows.	Ship Escort/Response Vessel System (SERVS) in place to help tankers navigate Prince William Sound and respond to spills or tankers in distress. In 1999, two new high-powered "tractor" tugs, custom built for the Sound, are put into service. Open-ocean rescue tug in place at Hinchinbrook Island.
Thirteen oil-skimming systems on hand with combined recovery capability of 1 million gallons in 72 hours.	More than 60 skimming vessels on hand with combined recovery capability of more than 12 million gallons in 72 hours.
Limited ship-tracking system by Coast Guard in Valdez.	Satellite-positioning data and radar permit Coast Guard to track tankers from Valdez terminal to Hinchinbrook Entrance, gateway to the Gulf of Alaska. Qualifications and training for watch standers are heightened. Repeater towers installed by Alyeska Pipeline Service Co. improve communication between tankers and the Valdez marine terminal.
Spill resources and equipment, detailed in contingency plans approved by state regulators, assume a "catastrophic" spill greater than 8.4 million gallons is highly unlikely. (The Exxon Valdez spill exceeded this amount by 2.6 million gallons.)	Revised regulations require gear immediately available to handle spills up to 12.6 million gallons within 72 hours.
Less than five miles of containment boom stowed for Prince William Sound response.	More than 34 miles of boom on hand as well as barges to hold recovered oil and water.
No major spill drills held; no plan to involve area fishing boats in a spill response.	Major drills held yearly; numerous smaller drills organized. Response plan includes fishing boats and vessel training.
Limited state oversight of Valdez marine terminal and tanker operations.	State funding increases significantly for spill drills, facility and vessel inspections and review of spill contingency plans submitted by oil shippers and Alyeska Pipeline. An oil spill response fund in place before the 1989 accident is expanded.
No established wildlife rescue programs.	Plan in place, including rescue and rehabilitation gear.
No protection plan for Prince William Sound fish hatcheries.	Plans established for Prince William Sound sites as well as an Afognak Island hatchery; pre-staged equipment on hand at hatcheries.
No oil slick dispersant and application systems detailed in contingency plan.	Dispersant is stockpiled; aircraft and ship-based application systems in place.
Minimum citizen oversight of tanker operations.	Citizen watchdog groups in place for Prince William Sound and Cook Inlet.

(Source: Prince William Sound Regional Citizens' Advisory Council; BP Exploration (Alaska) Inc.)

Post-Exxon Valdez regulations now address drug and alcohol testing despite a lack of evidence that alcohol consumed by Hazelwood in the hours before his ship set sail had significantly impaired the judgment of a master who, most colleagues acknowledged, was an excellent mariner. As a 1990 National Transportation Safety Board review determined, compounding error on the part of the ship's crew contributed to the accident. While drug and alcohol screenings foster overall safety, no one believes they alone can ensure against another disaster.

Exile as prevention was mandated too. A provision of OPA '90 banned the Exxon Valdez tanker from ever returning to Alaska waters. In the white-hot hours of the spill's aftermath, as Alaskans unfurled homemade, anti-big-oil banners or slapped pro-industry bumper stickers on their cars, the banishment of a tanker might have satisfied some. But risk managers thought otherwise. After repairs, the Exxon Valdez would remain one of the most advanced ships in the Alaska oil fleet; why blame the tanker?

The Oil Pollution Act also increased financial liability placed on tankers calling at U.S. ports. Congress hoped the prospect of paying more money following a spill would spur the industry to improve itself while also guaranteeing that, in the event of a large spill, a responsible party would make things right.

But experience has shown that this well-intentioned logic also falls short. The international tanker owners federation, based in London, warned Congress that OPA '90's financial liability mandates would prompt shippers to obscure responsibility behind a series of corporate curtains. In fact the economics of shipping oil have turned old ships into ones that produce the best profit margin; these ships now are registered in countries such as Liberia, which lacks a port big enough to hold all vessels registered there. Meanwhile tankers are staffed by crews not licensed in the United States, sometimes unable to speak the same language as their bridge commanders. In Cook Inlet, where liquid natural gas tankers call at the Kenai Peninsula port of Nikiski, tankers routinely have been staffed by foreign-based crews whose officers are from another country. The goal of improved communications among the crew — a key lesson from the Exxon Valdez wreck — has yet to be achieved.

Experts agree that insurance companies, and not regulators alone, wield power to bring about immediate improvements in ship safety. While regulations may be negotiated, there is little leverage when it comes to buying insurance since, if policy conditions are not met and an accident occurs, insurance may not be in force. In their book *Innocent Passage: The Wreck of the Tanker Braer* (1993), Jonathan Wills and Karen Warner note that the British commercial fleet has been driven offshore so that only a few hundred such vessels actually fly British colors. The port fleet calling on the Shetland Islands terminal at Sullom Voe makes up just 5 percent of the total British flag fleet. Despite OPA '90's stern liability mandates aimed at decreasing risk in U.S. ports, there still is world enough for irresponsible shippers to hide.

Technological improvements in tanker safety since 1989 abound: Radio repeaters now located in the Sound increase communication coverage while a sophisticated Vessel Traffic System permits Valdez-based Coast Guard to view and respond to even minute changes in a vessel's location. This upgraded system includes global positioning and a geographic information system to display locations. One stormy night within a few weeks of the system's initial testing, a Valdez Coast Guardsman who detected that a tanker had dragged its anchor a hundred feet or so went on to warn the ship's crew who had yet to notice anything amiss.

But miscommunication remains an obstacle: In 1995, the tanker Kenai narrowly missed running aground while navigating Valdez Narrows, a mile-wide choke point that separates Port Valdez from Prince William Sound. Investigations by the state revealed a series of miscommunications and failures to communicate among the vessel's bridge crew, the marine pilot on

board and the escort vessel as well as Coast Guardsmen using the Vessel Traffic System. The Kenai sailed to within a quarter-mile of the narrows' walls; catastrophe was avoided when the vessel slipped through and into the main part of the Sound.

Risk managers trying to encourage communication among crews and their superiors know they are bucking a centuries-old hierarchy that seldom rewards mariners who speak up. "Merely providing more equipment will not prevent accidents," the

National Research Council concluded in 1993, "improved training and education can prevent marine accidents."

In searching out improvements in oil spill prevention, credit must go to the oil industry itself in a new partnership with interest groups outside of government. At an oil spill conference held in March 1999 in Valdez, Alyeska president Bob Malone said preparedness had advanced in part as industry learned the value of working with others, including spill zone communities.

While exile of the Exxon Valdez had some popular support, expressed in this painting by Ketchikan's Ray Troll, authorities noted that the Valdez ranked among the newest tankers in Alaska's aging fleet. (Courtesy of Ray Troll)

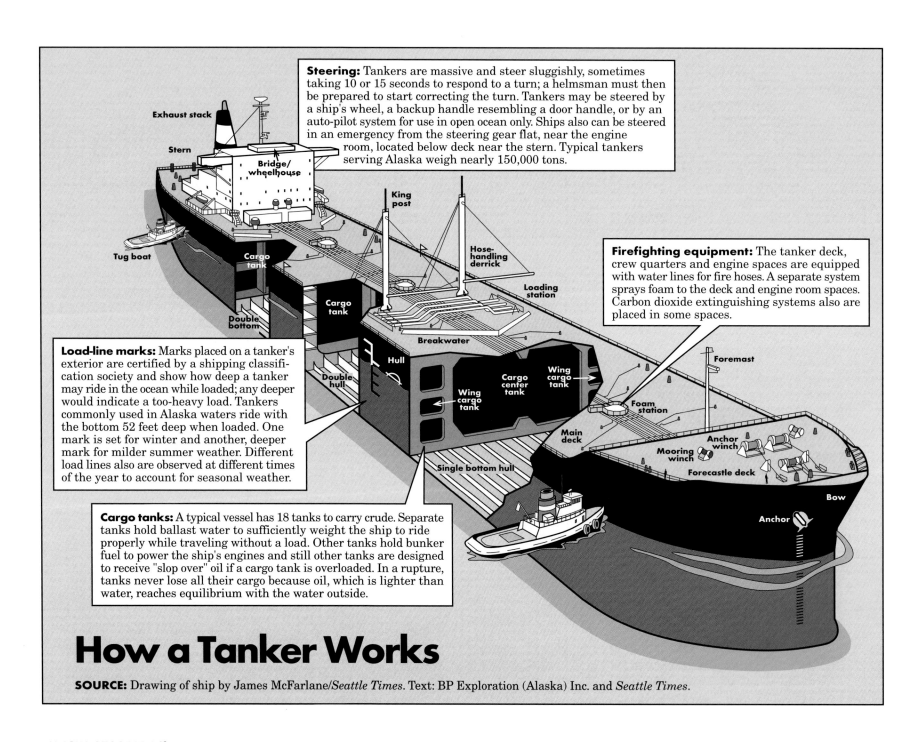

Steering: Tankers are massive and steer sluggishly, sometimes taking 10 or 15 seconds to respond to a turn; a helmsman must then be prepared to start correcting the turn. Tankers may be steered by a ship's wheel, a backup handle resembling a door handle, or by an auto-pilot system for use in open ocean only. Ships also can be steered in an emergency from the steering gear flat, near the engine room, located below deck near the stern. Typical tankers serving Alaska weigh nearly 150,000 tons.

Firefighting equipment: The tanker deck, crew quarters and engine spaces are equipped with water lines for fire hoses. A separate system sprays foam to the deck and engine room spaces. Carbon dioxide extinguishing systems also are placed in some spaces.

Load-line marks: Marks placed on a tanker's exterior are certified by a shipping classification society and show how deep a tanker may ride in the ocean while loaded; any deeper would indicate a too-heavy load. Tankers commonly used in Alaska waters ride with the bottom 52 feet deep when loaded. One mark is set for winter and another, deeper mark for milder summer weather. Different load lines also are observed at different times of the year to account for seasonal weather.

Cargo tanks: A typical vessel has 18 tanks to carry crude. Separate tanks hold ballast water to sufficiently weight the ship to ride properly while traveling without a load. Other tanks hold bunker fuel to power the ship's engines and still other tanks are designed to receive "slop over" oil if a cargo tank is overloaded. In a rupture, tanks never lose all their cargo because oil, which is lighter than water, reaches equilibrium with the water outside.

Exhaust stack
Stern
Bridge/wheelhouse
Tug boat
Cargo tank
Double bottom
King post
Hose-handling derrick
Loading station
Cargo tank
Breakwater
Hull
Double hull
Wing cargo tank
Cargo center tank
Wing cargo tank
Single bottom hull
Foremast
Foam station
Main deck
Mooring winch
Anchor winch
Forecastle deck
Bow
Anchor

How a Tanker Works

SOURCE: Drawing of ship by James McFarlane/*Seattle Times*. Text: BP Exploration (Alaska) Inc. and *Seattle Times*.

Eshamy Bay on the Kenai Peninsula's east side was among heavily polluted sites in 1989. Spilled oil drifted hundreds of miles on Gulf of Alaska currents. (Kevin Hartwell)

Since the Exxon Valdez wreck, oil shipping has been monitored by the Anchorage-based Prince William Sound Regional Citizens' Advisory Council, a watchdog group that traces its roots to OPA '90. "We value the input," Malone said at the March conference. "And while I don't always agree, I know RCAC has helped make our prevention and response efforts more effective."

Gov. Tony Knowles, who addressed the group as well, has said Alaska had heeded painful lessons from the tanker accident, which he called "the worst kind of spill, in the worst possible place."

"Alaskans have learned the price of complacency," Knowles said, noting such post-spill improvements as new escort tugs; enhanced training for tanker and tug officers and marine pilots; stockpiled spill gear; and citizen oversight. "We must be committed to doing it right."

Regulators, industry and oversight agencies have stressed that oil spill prevention is a system, not a series of

individual improvements or equipment additions. "We can never relax," the citizens' advisory council stated in its report prepared for the spill's 10th anniversary. "(The council) believes Alaska waters and communities affected by the Exxon spill are safer today. The oil industry operates with a heightened awareness of the consequences of a catastrophic spill. Continued vigilance is essential."

Debate over the need for "tractor" tugs in Alaska is a good example of vigilance as process. Longtime critics of oil spill preparedness in Alaska began looking to tractor tugs by the mid-1990s. Proponents said the highly maneuverable, more powerful tugs, capable of towing a tanker in stormy seas, could be a linchpin in enhancing tanker safety in the Sound. Tugs were in use all over the world, including some of the smallest ports; state officials wanted to know why some could not be stationed in Prince William Sound. In 1998, Alyeska Pipeline added new tractor tugs to its tanker escort fleet and in early 1999 christened another, the 10,000-horsepower Nanuq. The vessel was called into service within weeks, standing by as a tanker took on oil in Valdez in

FACING PAGE: *A Russian Orthodox church serves as a landmark in Tatitlek, along the Sound's northeast coast. Tatitlek is among Native villages that have sold lands for permanent conservation. (Patrick J. Endres)*

RIGHT: *Contingency planners seeking to learn from the Exxon Valdez wreck say a top priority is preventing spilled oil from reaching shore, where cleanup is slow, expensive and inefficient. (Al Grillo)*

winds to 50 mph. No losses were reported.

But like the double-bottom, double-hull issue, experts disagree about the extent of enhanced spill protection afforded by the new tugs. While acknowledging the tugs' place in preparedness planning, industry officials also caution that a piece of equipment on its own shouldn't be regarded as a linchpin. That conclusion was echoed in a two-year study of risk assessment in the Sound, completed in 1996 by the international marine casualty experts Det Norske Veritas, which favored investment in a variety of escort and response vessels that could be deployed at various sites in the Sound.

In place today are powerful escorts stationed at the entrance to the Sound,

a critical transition point from protected waters to the open ocean. Today's practice of escorts for inbound tankers without oil is noteworthy as well since oily ballast water carried by the vessels, coupled with tens of thousands of gallons of fuel and hydraulic oil on board, present a risk as well. This modified "sentinel" system is a joint product of shippers, government agencies and pressure by the citizens' advisory council; it was not specifically mandated by regulation or law.

Assessment by Det Norske Veritas noted that improvements through 1995 had reduced the risk of a tanker accident in Prince William Sound by 75 percent since 1989. The biggest gain was achieved when

LEFT: *The Sound is among few waterways where abundant wildlife, such as these sea lions heralding a tanker, coexist with commercial traffic. (Nick Jans)*

FACING PAGE: *Among provisions of the 1990 Oil Pollution Act: Banning the repaired Exxon Valdez tanker from sailing again in Alaska waters. (Roy Corral)*

limits in the Sound and more stringent tanker inspections, among numerous other mandates, the question now is how to eliminate remaining risk. Regulators, citizen watchdogs and the oil industry agree that improvement can be attained through continued talks coupled with technical analysis. Steady improvement in tanker safety also will be realized through enhanced mariner training which today includes bridge management classes that examine how decisions are made and communicated. Simulated training not unlike systems used by airline pilots is offered as well.

Gear and More Gear: Prevention's Lifeline

Anyone who lived through the Exxon Valdez oil spill cleanup understands why prevention today emphasizes gear: Two of the biggest problems faced by cleanup organizers in 1989 centered on equipment — not enough to pick up spilled oil and nowhere to store oil eventually collected. A third stumbling block was strategy for a major on-the-water recovery.

By 1999, industry and government had addressed all three hurdles. Today, regulations require 60 skimming systems on hand capable of collecting more than 12

the oil industry, at government urging, began the Ship Escort/Response Vessel System, or SERVS, calling for tug escorts for all loaded tankers passing out of the Sound. While the Coast Guard always has required loaded tankers to be under tug escort while traveling the Valdez Narrows, requirements today extend coverage through Hinchinbrook Entrance, gateway to the Gulf of Alaska. At least two escorts, including one tethered to a tanker's stern in the narrows, are mandated; three escorts may be required through the narrows in high winds. In Washington state, lawmakers used the spill's 10th anniversary to renew

calls for tugboat escorts for tankers transiting the Strait of Juan de Fuca.

Alyeska president Bob Malone, in comments to the Valdez conference, called SERVS the foundation of the company's spill prevention efforts; the advisory council regards the unit, the largest of its kind in the Western Hemisphere, as one of "the top oil spill response forces in the world." In addition to aiding a tanker in distress and initiating oil spill response, the escorts also are extra eyes to watch for early signs of trouble in transit.

In the post-Exxon Valdez days of breath tests for tanker crews, established speed

World Oil Spills

Here is a list of the world's 10 largest marine oil spills since 1989. The list notes vessel or incident, gallons lost and where.

1. Jan. 26, 1991: oil terminals, other installations and tankers are intentionally allowed to spill during Gulf War hostilities; 240 million gallons; off the Persian Gulf coast and Saudi Arabia

2. April 11, 1991: tanker Haven; 42 million gallons; Mediterranean Sea at Genoa, Italy

3. Jan. 5, 1993: tanker Braer; 25 million gallons; Garth Ness, Scotland.

4. Dec. 3, 1992: tanker Aegean Sea; 22 million gallons; La Coruna port, Spain

5. Feb. 15, 1996: tanker Sea Empress; 21 million gallons; Milford Haven harbor, England.

6. Dec. 19, 1989: tanker Khark; 20 million gallons; Atlantic Ocean off the Moroccan coast

7. May 28, 1991: tanker ABT Summer; 15 million gallons; Atlantic Ocean off Angola

8. (tied with No. 7) April 26, 1992: tanker Katina P.; 15 million gallons; Indian Ocean off South Africa

9. March 24, 1989: tanker Exxon Valdez; 11 million gallons; Prince William Sound

10. Oct. 21, 1994: tanker Thanassis A.; 10.9 million gallons; South China Sea off Hong Kong

NOTE: On a worldwide list of oil spills of 10 million gallons or more since 1960, the Exxon Valdez accident ranks 53rd out of 65. All but 10 of these spills are marine oil spills.

SOURCE: *Dagmar Schmidt Etkin, International Oil Spill Database and Oil Spill Intelligence Report, Arlington, Mass.*

Alaska Oil Spills

Here is a list of the top 10 Alaska spills by volume since the Exxon Valdez accident in 1989. The vessel or incident is followed by gallons lost, location and cause.

1. May 28, 1990: Alaska Railroad derailment; 100,000 gallons of diesel; roughly 20 miles north of Nenana.

2. Nov. 26, 1997: freighter Kuroshima; 39,000 gallons of bunker oil; grounded during a storm in Dutch Harbor.

3. Nov. 17, 1990: Little Diomede village tank farm; 31,000 gallons of diesel; storm erosion.

4. Aug. 10, 1994: contruction barge; 20,000 gallons of diesel; grounding at Cape Nome.

5. July 22, 1995: fish processor Northern Wind; 15,000 gallons of diesel; grounding at Nazan Bay in the Aleutian Islands.

6. Jan. 3, 1992: Port Nikiski pipeline rupture; 9,500 gallons of crude oil; Cook Inlet.

7. May, 21, 1994: tanker Eastern Lion; 8,400 gallons of crude oil; Valdez marine terminal; hull failure.

8. (tied with No. 7) Feb. 17, 1999: PetroStar refinery; 8,400 gallons of jet fuel; Valdez; overflow in storage tank.

9. June 26, 1997: city of Gambell tank farm; 8,000 gallons of diesel; leak.

10. Dec. 5, 1995: overfilled tank at Tesoro refinery at Nikiski; 5,700 gallons of crude oil; Cook Inlet.

SOURCE: *Camille Stephens, Alaska Department of Environmental Conservation.*

million gallons of oil-water mix in 72 hours. Compare that to 1989 when 13 systems, with a 72-hour capacity of about 1 million gallons, were available Just as important, barge capacity now on hand can store up to 34 million gallons of oil-water mixture recovered by skimmers. In 1989, during April and May, skimmers often were idle because storage barges were full.

Some 35 miles of containment "boom" (buoyant material used to corral a spill) are available today in the Sound, compared with less than five miles a decade ago. Escort-response vessels, themselves an addition since 1989, carry first-line response equipment, including spooled boom ready to drop in the water. New support units such as a dock, warehouse, storage space and fuel are within relatively easy reach at Tatitlek, 10 miles east of Bligh Reef, and at Chenega Bay, roughly 50 miles southwest of Bligh. Response gear is cached at strategic locations elsewhere around the Sound.

Packages of chemical dispersant are ready to install on aircraft along with 2,600 feet of fire-resistant boom and special ignition units to be dropped from helicopters, in case initial cleanup calls for burning oil on water. Fishing fleets in Valdez, Cordova and Kodiak have been trained, contracts in hand, to offer immediate support in a large spill — a task that took some time in 1989.

The Ship Escort/Response Vessel System employs nearly 200 specialists, although the unit has seen some decline in numbers with the drop in world oil prices. Another 60 people make up Alyeska's oil spill crisis management team. Under the direction of Anchorage-based BP Exploration (Alaska) Inc., the state's major oil producers used the 1998 response drill to test industry's ability to fly in extra equipment, including some arriving on short notice from as far away as England.

Ten years after the Exxon Valdez spill, only the most skeptical critic continues to insist that Alaska's oil producers have failed to mobilize gear and support needed to respond to another large accident. Regulations stemming from Alaska's experience have played a role; spillers today must operate within certain time and capacity limits. But change in Prince William Sound also is an example of a large industry marshaling its brightest and best to solve a problem and, mostly, getting it right. "We have all learned prevention is the key," Alyeska's Malone said.

Cleaning Up Oil, Clarifying Tasks

No list of lessons learned from the Exxon Valdez spill would be complete without mention of the Incident Command System, an easily overlooked advancement over the days, only a decade ago, when one of the first questions to be resolved after oil hit the water was just who's in charge.

Anyone present during those early hours in Valdez will recall the chaos that is the soul of a large oil spill. For everyone else, the confusion may be hard to appreciate since what's shown on television is a composed official explaining what's been mobilized where. Organizing a response to a large-scale spill in remote Alaska waters required hundreds of people and pieces of equipment spread over miles of uninhabited terrain that posed its own workplace hazards. Crews had to be accounted for, housed, fed and safely returned home; equipment required fuel, servicing and replacing; waste, industrial and human, had to be disposed of properly. Add concerned community leaders, elected officials and journalists from around the world and events spun faster day by day.

Eventually, the command system became the management scheme for authorities responding to the Prince William Sound

spill. The system rapidly clarified and centralized decision-making; disparate groups, some of whom would rather not be sitting around a table together, concentrated on a priority list so that cleanup could progress.

The system called for a field commander and policy-making team that included the spiller, the Coast Guard and Alaska state officials. Field objectives, such as which sites would be cleaned and how, were determined jointly so work could be done quickly with limited resources. Logistics and safety were bundled together, reducing the risk of worker injury.

As someone who participated in the command system, representing the governor's office, I know that meetings could be difficult. I also learned that the system worked by keeping all parties focused, making it harder for a side to opt out or play renegade. The command system fostered reasonable compromise in the heat of chaos and, if there is a next time, all sides now understand the management rules and procedures.

While a management system now exists, practice drills in Prince William Sound show a developing strategy when it comes to cleanup.

In 1989, state authorities insisted that mechanical recovery such as skimming from boats occur before chemical dispersants or other strategies were tried. The idea, state officials reasoned, was to remove as much oil from the environment before treating

The Debate Over Double Hulls

Among spill-prevention strategies contained in OPA '90 is a mandate for double-hulled tankers by 2015, a requirement that consumed policy makers, engineers and regulators for years following the Exxon Valdez accident.

The goal is to reduce cargo loss by designing a "tank within a hull," offering two steel barriers and a 10-foot void space between the sea and oil. Double hulls, in which a second skin of metal is run along a tanker's sides, offer greater protection in case of collision; double bottoms can protect in a grounding. Critics of double hulls have focused on alleged inspection difficulties, risk of fire and explosion and claims that double-bottoms are unstable during salvage if they take on water.

Proponents, meanwhile, favor incentives to speed the construction of double hull tankers and barges. A study, *The Double Hull Issue and Oil Spill Risk on the Pacific West Coast* (1995), prepared for the British Columbia government notes that groundings account for more than half the volume of oil spilled from U.S. tankers. "Two large independent tanker associations (have) stated that double hulls, although more difficult to maintain, can still be operated in a satisfactory manner," the report said.

While Coast Guard estimates show Alaska's spill would have been at least 60 percent smaller had the ship been fitted with a double hull, skeptics note the grounding was hard enough to overcome two barriers. Writer Ernie Piper, who glimpsed the gashed hull, agrees: "I'm not convinced," he says, "that two skins would have made much difference when a fully loaded tanker running at 14 mph rams a 30-million-year-old rock."

To date, BP Oil Shipping Co. has three double-hulled tankers and six double-bottomed tankers assigned to the Sound; Arco Marine has announced plans for double-hull tankers for Alaska as well. •

with chemicals that carry their own potential for harm

But all sides now agree that time, not technique, is the critical element in a successful cleanup. Containing spilled oil and removing it from the water or shore is very inefficient and its success depends on weather and luck, not the quantity or quality of equipment on hand. After oil has weathered on the water's surface even a few dozen hours, every cleanup technique starts hitting the edges of its effectiveness.

The inescapable problem is that oil does indeed mix with water. The sludgy emulsion, known as "mousse," does not skim easily because it is too cold and thick and may trap debris such as sticks or bits of buoys. Meanwhile, oil-on-water burning, a great hope of spill responders over the past

Both industry and citizen watchdog groups credit the use of ship escort/response vessels for enhancing safety since 1989. (Al Grillo)

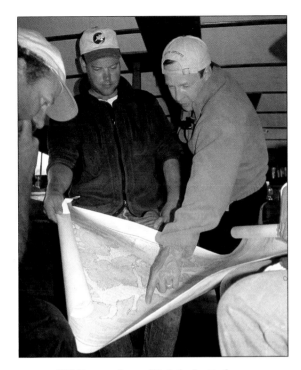

ABOVE: *While tourism officials fretted over images of oiled Alaska, the Sound's appeal to boaters, anglers and wildlife watchers remains strong. Ten years after the spill, tourists book cruises with Seattle-based "Earth Odysseys" to explore Prince William Sound while learning about the environment. The company's senior guide, David Sale (center), holds a navigation chart. (Beth Whitman)*

RIGHT: *As state on-scene coordinator, author Ernie Piper surveys an oil spill site in 1993. (Anchorage Daily News)*

10 years, remains problematic: Spilled oil must be corralled into discrete slugs and allowed to accumulate so that crude oil's volatile gasses will catch fire. Actually igniting the slick is a feat for helicopter crews; fire must then be monitored to make sure it is contained and boom has not melted.

Chemical dispersants have a small window of effectiveness as well. After two large practice drills in Prince William Sound, most observers now believe that the top priority in a spill's early days is to keep oil from washing ashore, where it is even harder to make good decisions about retrieval. If there is a next time in Prince William Sound, authorities are more likely to try a mix of oil-removing techniques simultaneously — burning where that's reasonable, skimming elsewhere and dropping dispersants on leading edges to keep the slick from reaching coastline.

Leave It to Nature?

Along with images of Alaska's oiled birds and beaches, the spill's 10th anniversary also revives difficult debate over how, when and where to clean oil from once-pristine coast. No one doubts that if oil hits the shores again, cleanup will be ordered but the debate will be no easier despite hard lessons learned.

One of those came midway through the summer of 1989 when federal and state scientists began to publicly question the efficacy and long-term effects of high-pressure, hot water washing that had dominated that season's work. Scientists concluded that shoreline organisms and ecological systems were better off awaiting nature's scouring. Removing what was classified as "gross contamination," where thick oil covers more than 50 percent of a shoreline area, might be worthwhile, these experts acknowledged. But they said once that was accomplished, wind and waves, along with the natural but slow breakdown of oil ultimately would leave the shoreline roughly the way it once was.

Since those findings were announced, studies including work by Allen Mearns of the National Oceanic and Atmospheric Administration generally support the idea. In 1989, still other experts acknowledged

FACING PAGE: *"Nature's resiliency, combined with cleanup efforts, has restored this vast and valued resource," Exxon noted on the spill's 10th anniversary. Said the oil spill trustees: "There is a long way to go." Crews practice maneuvers with absorbent boom in Valdez harbor two weeks after the tanker grounding in 1989. (Alissa Crandall)*

that hot water washing disrupted the environment, but so did a two-inch ring of oil. Also weighing in was the state, which insisted that interests such as fishing and recreation sometimes could outweigh the pure biological value of leaving an area to clean on its own.

Today there is abundant scientific data to support a leave-it-to-nature approach to oil spill cleanup; there is only anecdotal, non-technical information in favor of aggressive cleanup because economics or recreation deserve consideration too. Without more study by public agencies, this consideration could fade as cleanup plans are laid if there is a next time.

This tension between the dictates of science and the public's desire to restore nature has led to work on devising a general standard for all shoreline spills, a development that traces its roots to the Exxon Valdez spill. Leaders in this effort include Jacqui Michel, a lead consultant to NOAA and others, and industry consultant Ed Owens, a highly experienced Canadian spill expert.

Much of Owens' and Michel's work is aimed at standardizing shoreline observations, giving decision-makers comparable pieces of information and eliminating politics and subjectivity when it comes to scheduling cleanup work. Owens has even developed a software application that helps steer this process. While experts such as engineers and biologists still resist assigning much cleanup discretion to lay people, interest groups on all sides may welcome efforts to inject a degree of precision into the highly inexact science of assessing oiled shorelines. The issue was among panel discussions held in Seattle in 1999

at the International Oil Spill Conference.

It was hard in 1989 to show that diversions such as fishing or kayaking deserved consideration as cleanup options were weighed and either put in place or discarded forever. Not all recreational or public concerns will rival the interests of science, but some do; if there is a next time, they again warrant consideration as much as the health and well-being of limpets and algae.

But that's for the next round of cleanup managers to resolve, probably on a case-by-case basis and almost certainly in a less-contentious climate than the one that developed in 1989. Ten years after the Exxon Valdez wreck put tiny Prince William Sound on the international agenda, technology still lags behind needs, laws have their limits and plans are only paper after all. As the past decade proves, the key to oil spill prevention is people, starting with the meticulous mariner, the conscientious chief executive officer and the government official who knows how to solve a problem, not whom to sue.

In these respects, it's still March 24, 1989, and will always be. ✈

As a special assistant to the governor in 1989, Ernie Piper was focusing on resource issues when the Exxon Valdez oil spill occurred, moving him into position as the state's liaison with displaced communities. He later managed state agency and community response programs as Alaska's on-scene coordinator; in 1993, Piper prepared the state's final report on the accident. An ALASKA GEOGRAPHIC® *author and former Anchorage journalist, Piper today is public affairs director for the Alaska Railroad.*

When Oil Hits Water: A Spill Chronology

1989

MARCH 24: The oil tanker Exxon Valdez runs aground on charted Bligh Reef in eastern Prince William Sound, about 25 miles south of the trans-Alaska oil pipeline terminal at Valdez. Experts immediately warn that the 11-million-gallon spill occurs as fish, marine mammals and birds are headed for the Sound in spring migration. Eight of the ship's cargo tanks rupture, causing an estimated $25 million in damage to the ship; lost oil is valued at more than $3 million. No injuries were reported.

MARCH 25: Authorities approve a burn of oil floating on water, a preferred technique when spills are fresh. Some 15,000 gallons are reduced to tarry residue.

MARCH 25: State estimates show the spill covers 2,500 acres at an average thickness of about one-tenth of an inch. An initial test of chemical dispersant fails when winds are too calm to generate sufficient mixing. Dispersants are not tried again after winds intensify and the pancakelike spill begins to break up.

MARCH 26: Two days of unusually calm weather end and high winds further disperse oil. Burning is not attempted again when authorities note that oil had been whipped to "mousse," containing as much as 80 percent water.

MARCH 26: Gov. Steve Cowper declares a state of emergency, clearing the way for use of Alaska National Guard equipment, crews and aircraft to combat the slick.

MARCH 27: Oil is carried west on Gulf of Alaska currents; aerial tracking produces the state's first map of the spill's size and movement. More than 260 boats are dispatched to skim oil.

MARCH 27: Seeking to deflect oil headed for a western Prince William Sound salmon hatchery, Cordova-based fishermen join state and military crews to protect Sawmill Bay. Several dozen volunteers and about 40 fishing boats successfully lay absorbent boom and scoop oil in five-gallon buckets, hauling up as much as 1,000 barrels a day of oil-water mix.

MARCH 30: A sea otter cleaning and rehabilitation site opens in Valdez. The first bird rehabilitation site opens March 31 in Valdez; other cities are added as oil moves south.

EARLY APRIL: Oil washes ashore over large areas and shoreline cleanup in Prince William Sound and the Gulf of Alaska begins. Some crews hand-wipe oil from rocks, an effort quickly abandoned in light of the spill's size. At its peak, 10,000 workers are hired for cleanup in 1989. Exxon Corp. announces plan to complete the task by Sept. 15.

APRIL: Valdez's population quadruples to 10,000 as the town is transformed into oil spill headquarters. Officials hold daily public briefings and mobilize what eventually will become a $2 billion cleanup lasting nearly four years. Journalists arrive from around the world.

APRIL: State officials claim the industry failed to store sufficient spill-response equipment at Valdez as required in emergency

BELOW LEFT: *Hardest hit by the slick in 1989 were birds, which swam through oil or ingested it during preening. Treatment centers to wash oil from feathers reported only mixed success after birds were returned to the wild. (L.J. Evans)*

BELOW: *News of Alaska's 11-million-gallon spill in 1989 made headlines worldwide; intervening years brought steady reports of multimillion-dollar cleanup efforts, lawsuits and land purchases. In 1999, international news teams fanned out again in the spill zone to document conditions. (Exxon Valdez Oil Spill Trustees)*

plans; critics fault state regulators as well, saying Alaska's spill-response plan was inadequate. Alaska begins hearing from entrepreneurs suggesting crushed cork, lemon juice and other novel methods to remove oil from water.

APRIL: Coast Guard Vice Adm. Clyde Robbins, then federal on-scene coordinator, sets in place a management structure adhered to for the next four years of cleanup: Exxon cleanup plans

BELOW: *Coast Guard authorities praised cleanup crews in 1989 while also concluding the spill zone required more work. Oil is trapped in a cobble beach near Seward on the Kenai Peninsula, 120 miles southwest of the tanker grounding. (Exxon Valdez Oil Spill Trustees)*

BELOW RIGHT: *Hot water is sprayed from a high-pressure hose to flush oil from rocky beaches. Scientists worried however that hosing left behind sterile beaches, and suggested instead that some oiled sites be left for nature to repair. (Al Grillo)*

are submitted to state agencies and other landowners for comment; state and federal authorities may then alter the proposal before authorizing Exxon to begin work. Government officials verify that work is completed satisfactorily.

APRIL 3: Innovative "supersucker" techniques emerge to remove some 450,000 gallons of weathered, oil-and-water mousse. The Army Corps of Engineers uses a dredge with suction units turned upside down while from North Slope oil fields, two large vacuum trucks are driven 800 miles to tidewater under highspeed trooper escort.

APRIL 3: The first in a series of fishery closings extending from

Cordova to Chignik on the Alaska Peninsula is ordered as the state shuts down the Prince William Sound herring fishery to successfully keep tainted seafood from reaching market. Fishermen eventually are idled in Cook Inlet and Kodiak as well, and Exxon establishes a voluntary claims system to compensate the fleets. Some fishermen contract with the company to take on lucrative cleanup support work.

APRIL 5: After lightering is begun March 25 to remove remaining oil, the 987-foot Exxon Valdez is refloated at high tide, temporarily patched and anchored off Naked Island in Prince William Sound. The tanker, built in 1986 and among the newest in Exxon's fleet, eventually is towed to drydock in San Diego beginning June 20.

APRIL 19: At a hearing before the U.S. Senate Subcommittee on Environmental Protection, Alaska Gov. Steve Cowper calls for a nationwide program on oil spill response: "We should not have to use a spill like that in Prince William Sound to find out the best way."

APRIL 21: Exxon contracts with the world's largest available skimmer boat, the 425-foot Soviet Vaydaghupski, but it arrives after weather deteriorates.

APRIL 26: Oil is seen washing from Smith, Little Smith and Seal islands in western Prince William

Sound and Point Eleanor, areas heavily hit in the initial release. "Reoiling" is identified as a problem and officials pinpoint Point Helen, at Knight Island's southern tip, and Sleepy Bay, at LaTouche Island's north end, as most heavily oiled sites. Both Knight and LaTouche islands will see cleanup into 1992.

MAY 5: Vice President Dan Quayle makes a brief inspection of Smith Island and meets in Anchorage with mayors of oil spill displaced towns.

MAY 16: A federal inquiry into the tanker wreck by the National Transportation Safety Board begins in Anchorage and concludes May 20. In an accident report published in July, the board notes several probable causes including a third mate's maneuver error, which the NTSB blamed on fatigue and excessive workload. Other safety factors named by the board: the skipper's "impairment" from alcohol and a lack of effective vessel traffic control by the Coast Guard at Valdez.

MAY 18: Oil is detected along the Alaska Peninsula, the spill's farthest reach, 470 miles west of Bligh Reef.

JUNE: The Pratt Museum in Homer unveils "Darkened Waters," an exhibit focusing on the Exxon Valdez oil spill and cleanup as well as Alaska's oil history. A traveling version eventually was displayed at 15 sites nationwide, including the Smithsonian Institution in 1991.

AUG. 14: The state sues Alyeska Pipeline Service Co. and Exxon for negligence and unspecified damage to the environment; Exxon countersues on Oct. 24, claiming state interference in cleanup plans.

SEPT. 15: In Valdez, Coast Guard Commandant Paul Yost praises Exxon's cleanup crews but notes "they didn't finish the job," and requires Exxon to resume cleanup in spring. The state Department of Environmental Conservation and Exxon classify 200 miles of shoreline as heavily or moderately oiled; 1,300 miles are oiled overall.

1989: Exxon pays out-of-court settlements of about $130 million to commercial fishermen, cannery workers and some local governments.

1989: The Alaska Legislature responds to the tanker accident with reforms including increased civil fines for oil spills; tax law changes that bar Exxon and Alyeska Pipeline Service Co. from deducting cleanup costs from state severance taxes; and a revision to contingency plan requirements.

1989: The state's subsistence foods division notes that total pounds collected by Alaska Natives in the Prince William Sound villages of Chenega Bay

and Tatitlek had dropped by more than half. Testing in 1990 and 1991 eventually concludes that traditional foods such as seal, salmon, deer and gull eggs are safe as long as they do not look or smell oily.

1990

FEBRUARY: The state's Oil Spill Commission, formed to examine the tanker accident, concludes that spill response technology is largely untested and underdeveloped. Reviewers also note that state contingency plans had failed to envision a spill of the Exxon Valdez's size.

FEBRUARY: Federal, state and Exxon officials begin planning the 1990 cleanup season. Among techniques reviewed is "bioremediation," the use of fertilizers to enhance natural breakdown of oil in the environment. Dissension develops when Alaska favors aggressive cleanup while National Oceanographic and Atmospheric Administration scientists propose leaving oil unless a compelling reason can be shown to remove it. Moreharm-than-good concerns arise about oil-removal techniques such as excavation.

MARCH: Winter storms are credited with scouring spilled oil, reducing its overall presence by almost 20 percent. Estimates of heavily oiled shoreline drop from an aggregate of 53 miles to just 15 miles, nearly all in Prince William

Sound. While acknowledging improvement, the state declares remaining pollution "massive."

MARCH 23: Exxon Valdez skipper Joseph Hazelwood, fired after the grounding, is acquitted in Anchorage Superior Court of operating a vessel while drunk. Jurors convict Hazelwood of one misdemeanor criminal count of negligently discharging oil and he is ordered to serve 1,000 hours of unpaid community service, including cleaning oily shoreline. After appeals, Hazelwood, who works today as a maritime consultant for his New York defense team, eventually was ordered to begin work service in Anchorage in 1999.

APRIL: Federal coordinators approve Exxon's 1990 cleanup

ABOVE LEFT: *Convicted in 1990 of negligently discharging oil, former Exxon skipper Joseph Hazelwood was ordered to clean oily rocks as part of his misdemeanor sentence. That demand was set aside in 1998 when, after lengthy appeals, Hazelwood was ordered to begin 1,000 hours of unpaid work service in Anchorage and pay a $50,000 fine (Cary Anderson)*

ABOVE: *Web of life: Oil-spill science is uncovering the interdependence of marine life by seeing how disruptions in certain species trickle down to others. In Prince William Sound, fucus and other aquatic plants offer shelter to fish. (Exxon Valdez Oil Spill Trustees)*

plan, which calls for about 200 workers relying mostly on hand tools and bioremediation fertilizer to remove oil.

APRIL 16: The state agrees to widescale use of fertilizers despite

ABOVE: *Knight Island in western Prince William Sound was among worst-case sites in 1989 and today harbors pockets of Exxon Valdez oil. Part of the island has been set aside as protected habitat for wildlife displaced by the spill. (Patrick J. Endres)*

ABOVE RIGHT: *Fairbanks researcher Gail Blundell cradles a sedated river otter tested for exposure to oil pollution in Prince William Sound. Declared a recovered species in 1999, river otters were little studied in the Sound before the spill. (Exxon Valdez Oil Spill Trustees)*

concern that they are ineffective against stubborn oil. State reviewers eventually conclude that bioremediation enhanced oil breakdown although efficacy was uneven.

MAY: Cleanup begins for a second season. Officials note that beach surveys completed earlier in the

year had missed asphalt patches and mousse hidden by snow.

JUNE: A technical advisory group, aimed at fostering consensus in evaluating cleanup work, encounters a snag when the Alaska Department of Fish and Game announces dissatisfaction, claiming the group fails to put proper emphasis on performance criteria. By 1991, the state would write its own work orders and do supplemental cleanup.

JULY: Following repairs, the Exxon Valdez tanker is renamed SeaRiver Mediterranean and moved to Exxon service in Europe.

SUMMER: Cleanup resumes with heavy equipment such as backhoes, tractors and front-end loaders to excavate large tar and

asphalt patches from beaches. Critics wonder if tilling creates problems by reintroducing oil into the water.

1990: The Coast Guard announces Exxon will return for a third cleanup season in 1991.

1990: Congress adopts the Oil Pollution Act, making available up to $1 billion per spill for response and waste removable costs. The act increases liability for oil owners and shippers, sets up a new oil pollution response fund for prevention and research, and requires stronger preventive measures by government and private parties. The bill also authorizes creation of regional citizen advisory councils, one as an oil shipping watchdog in Prince William Sound and the other for Cook Inlet.

1990: Alaska's Legislature adopts several spill-influenced bills that, among other things, give the Alaska Department of Environmental Conservation authority to inspect tankers and establish penalties for certain environmental crimes. Heeding Alaska's experience, California adopts a comprehensive oil spill prevention and response act.

1991

APRIL: In Washington, NOAA officials announce that hot-water, high-pressure washes used in Alaska may have caused environmental harm: "Sometimes the

best thing to do in an oil spill is nothing," the scientists conclude. State officials eventually declare the treatment an acceptable alternative at heaviest oiled sites.

MAY: State spill managers begin surveying some 600 shoreline sites, most of them in Prince William Sound, in preparation for the summer work season. Crews find that places heavily oiled in 1989 still show significant patches of surface oil as well as pockets of subsurface oil. In all, more than 17 miles of shoreline are identified as holding subsurface oil.

MAY THROUGH JULY: Officials recommend no treatment at nearly 500 sites identified in May. Cleanup crews, working mostly with shovels and rakes, remove roughly 700 tons of oily sediment from remaining shorelines. Bioremediation plays only a minor role as weathered oil is not responsive to chemical treatment.

JULY: Coast Guard science advisers suggest that high-pressure washing with water as hot as 160 degrees is taking a toll on certain marine life. The technique enters its sixth week.

OCT. 9: A landmark $1 billion agreement is approved to settle state and federal oil spill claims against Exxon. The agreement clears the way for Exxon to pay $900 million to the state and federal governments over a 10-year period to restore Alaska's environment. Terms also call for the company and its shipping unit to plead guilty to federal crimes against the environment, drawing a $150 million fine. State and federal agencies form a six-person trustee council to devise a restoration spending plan drawing on the civil settlement fund.

1992

WINTER: A final shoreline assessment is planned, targeting about 60 sites for survey in spring.

MAY THROUGH JUNE: Cleanup crews remove sediment and break apart small patches of asphalt at a few dozen sites.

JUNE 12: State and federal regulators declare the cleanup complete based on findings that

much of the spill area is free of oil. Trapped oil buried in rocky beaches or under mussel beds persists as does some asphalt, probably inert, at the surface.

1993

APRIL: Prince William Sound's $12 million Pacific herring fishery collapses when fish that hatched in 1989 fail to arrive on cycle as spawning adults in 1993. A latent

BELOW LEFT: *Shoreline assessment in 1993 included tests of chemical solvents. A worker displays the potency of one cleaner applied to an oily pad in Prince William Sound. (Natalie B. Fobes)*

BELOW: *Oil spill studies on pink salmon expanded the use of an innovative technique to "mark" the otoliths, or ear bones, of hatchery reared fish, to distinguish them from wild salmon. The process exposes developing fish to warmer waters, causing a harmless but detectable alteration of the otolith. The technique is being used in Alaska fishery management. (Exxon Valdez Oil Spill Trustees)*

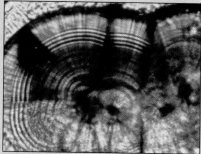

virus and fungus are blamed; researchers conclude oil may have caused the virus to spread.

AUGUST: At a cost of $22 million, the state acquires 24,000 acres of private land scheduled to be logged within Kachemak Bay State Park near Homer. The acquisition is the first major habitat protection effort following the spill.

1994

MARCH 22: In observance of the 5-year anniversary of the tanker wreck, oil spill scientists and restoration experts present findings to the public at an Anchorage conference. Analysts note Exxon stock has risen nearly 43 percent since 1989 while nationwide, theaters begin screening "On Deadly Ground," an action movie pitting an environmentalist against the oil industry.

SPRING: Researchers begin a 5-year, $21 million project to better understand plankton health and its role in the survival of juvenile herring and salmon. The Sound Ecosystem Assessment study, funded by the spill trustees, could eventually help predict fish abundance.

MAY: Afognak Island State Park, covering more than 41,000 acres of old-growth spruce forest, is created when $39.5 million in settlement funds are used to buy the parcel as cover for bald eagles, marbled murrelets and salmon.

SEPT. 16: After a four-month civil trial, an Anchorage federal jury hearing a class-action oil spill lawsuit orders Exxon Corp. to pay $5 billion in punitive damages. Jurors also find the company liable for $20 million in actual damages and, on June 13, found former Exxon Valdez skipper Joseph Hazelwood liable for recklessness. Exxon appeals, sending the case to the 9th U.S. Circuit Court of Appeals in San Francisco where a decision was pending in early 1999.

SEPT. 24: An Anchorage Superior Court jury awards nearly $10 million to six Alaska Native groups and the Kodiak Island Borough following a three-month trial over oil-damaged lands. Claimants sought more than $120 million from Exxon and in early 1999 were awaiting an appellate ruling from the Alaska Supreme Court.

NOV. 2: After 18 months of planning, Exxon Valdez oil spill trustees adopt a long-range environmental restoration program and a 10-year budget. The plan identifies species injured by the spill and specifies lands needed to spur recovery.

1994: Cleanup crews focus on 12 mussel beds in Chenega Bay to see if removing and replacing oily sediment under the beds will hasten restoration; 38 tons of oiled sediment are removed. Oil concentration dropped by an average of 98 percent by 1995, expert say, while mussel survival rates generally were good.

1994: A project to restore littleneck clam harvests in the oil spill region begins at Seward's Qutekcak hatchery. Operating from a $3 million mariculture unit, the project brings clam seedings to Chenega Bay and Port Graham villages.

1995

MAY: Kodiak's Alutiiq Museum, housing local archaeological artifacts, opens. The museum receives $1.5 million from spill trustees seeking to protect archaeological sites in the spill zone.

SPRING: Prince William Sound public school students learn about habitat restoration through Youth Area Watch, an ongoing program that places students with oil spill scientists in the field or laboratory.

AUGUST: Two species of birds, Kittlitz's murrelets and common loons, are added to the spill trustees' list of injured species.

NOVEMBER: Oil spill trustees agree to spend $89 million for nearly 270,000 acres within Kodiak National Wildlife Refuge to enhance habitat for salmon, bald eagles and sea birds.

DECEMBER: Land purchases by the oil spill trustees and the state quadruple the size of Shuyak Island State Park near Kodiak to nearly 50,000 acres.

1995: Research shows that sea otters in the spill zone are reproducing at about the same rate as sea otters outside the zone.

1995: Prince William Sound communities of Chenega Bay, Cordova, Tatitlek, Whittier and Valdez join in a regionwide project to reduce marine pollution. The effort eventually expands to include Kodiak Island and remote Cook Inlet towns.

1996

JUNE: Oil spill trustees pursue their first acquisition within Prince William Sound, offering to buy old-growth forest and 190 miles of coastline owned by Native-run Chenega Corp. The $34 million purchase is completed in 1997 and takes in 59,500 acres. The western Prince William Sound parcel includes salmon-rearing streams and nesting trees used by the marbled murrelet.

JULY: Oil spill research into genetic identification of sockeye salmon is applied on the Kenai River, where the technique is used for the first time to help manage a commercial fishery.

AUGUST: Oil spill trustees seeking to protect nearly 70,000 acres within Prince William Sound enter into an agreement with Native-owned Tatitlek Corp. The $34.5 million purchase is completed in October 1998.

SUMMER: Oil-spill funded research into long-term declines of harbor seals, pigeon guillemots and marbled murrelets suggests a drop in abundance of fat-rich forage fish may be a factor. Known as the Alaska Predator Ecosystem Experiment, the study indicates that warming waters in the Gulf of Alaska may be to blame for an increase in lean fish, such as pollock and cod.

SEPTEMBER: Based on findings from science advisers, the trustee

Commercial fishermen in 1997 try their luck in Nuka Bay, near Kenai Fjords National Park and among oiled sites in 1989. A decade after the spill, research into residual oil and its potential affect on fish continues. (Roy Corral)

council declares the spill zone's bald eagle population recovered from oil spill losses. More than two dozen species of birds, animals and fish are deemed still injured by the spill.

1996: A group of male sea otters returns to northern Knight Island, generating hope for recolonization at one of the Sound's most heavily oiled sites. The group fails to return in 1997 but in 1998, researchers announce that sea otters in the spill zone generally are recovering well.

1996: Scientists note a lag in recovery of brown seaweed known as fucus, or popweed, in the intertidal zone where shellfish and small invertebrates live. A critical link in the ecosystem, the seaweed provides protective canopy for species preyed on by otters and ducks, among others.

1996: Traditional hunters are trained in tissue sample collection to aid biologists evaluating the health of seals and otters in the spill zone.

1997

MAY 1997: Kenai Fjords National Park inholdings, totaling roughly 32,500 acres, are acquired for habitat protection at a price of $15 million.

JULY: Native-owned Eyak Corp. agrees to sell 75,000 acres in eastern Prince William Sound for habitat restoration. The tract, which sold for $45 million, includes roughly 80 salmon streams.

SUMMER: At villagers' urging, crews remove entrenched oil along Chenega beaches in western Prince William Sound.

SUMMER: Murres return to Barren Islands, an oiled site north of Kodiak Island, for the first

time after a large brood hatched in 1993. Researchers announce a setback in 1998 after a large die-off in summer that may be linked to the warm ocean current known as El Nino. About three-fourths of the 250,000 sea birds killed by the spill were murres.

AUGUST: Researchers use the Prince William Sound pink salmon fishery to test an innovative technique to distinguish wild and hatchery reared fish. Otolith marking, used for the first time in Alaska as an outgrowth of spill science, leaves a distinctive "tree ring" mark on the ear bone of hatchery salmon and can be readily detected in the field. The mark is generated naturally when water temperature is increased at hatchery incubators.

1998

APRIL 18: Exxon lawyers unsuccessfully ask an Anchorage federal court to allow the Exxon Valdez tanker, renamed SeaRiver Mediterranean, to return to Alaska. Court approval is needed after the federal Oil Pollution Act

Fishermen take to their boats in frustration over failed Prince William Sound harvests. Pink salmon and Pacific herring, the Sound's money fish, remain on a list of species injured by the spill. Ten years after the accident, disputes continue over how — or if — Exxon Valdez oil factors into fishing losses. (Al Grillo)

of 1990 ban-ned the ship from Alaska waters.

MAY: The Alaska SeaLife Center, a combined public aquarium and marine research laboratory, formally opens. The $55 million project is funded through the trustee council and an oil spill settlement controlled by state legislators.

NOVEMBER: Former Exxon Valdez skipper Joseph Hazelwood, convicted in 1990 of negligently discharging oil, is ordered to begin paying a $50,000 misdemeanor fine and begin work service. Hazelwood, who lives in New York, agrees to return to Anchorage one month each summer over the next five years to clean roadways and parks.

1998: The Nearshore Vertebrate Predator study begins examining oil's potential role in population declines that also may be linked to availability of food.

1998: Prince William Sound's well-studied "AB" pod of resident killer whales continues to show signs of stress after losing 13 of its 36 members within two years of the spill. In the Gulf of Alaska, however, resident killer whales increase from 102 animals in 1988 to 110 animals.

1998: No new vandalism is reported at archaeological sites damaged or looted during oil spill shore cleanup in the early

1990s. The trustee council sets aside nearly $3 million for new archaeological centers serving eight communities touched by the spill.

1999

FEB. 9: Oil spill trustees declare the spill zone's river otter population "recovered" and move several other species — clams, Pacific herring, sea otters, black oystercatchers and marbled murrelets — to the "recovering" list. By early 1999, river otters and bald eagles are the only two of nearly two dozen injured species deemed recovered from the accident.

MARCH: Scientists, elected officials, environmentalists and the oil industry gather at several conferences observing the accident's 10th anniversary. Sessions focusing on the status of Alaska's spill zone and progress in shipping oil safely are held in Seattle, Valdez and Anchorage

MARCH 1: Exxon Valdez oil spill trustees deciding how to spend roughly $170 million in reserve funds from Alaska's civil settle-ment vote to set aside $115 million for long-term research, monitoring and community projects and $55 million for continued habitat protection efforts. The vote con-tinues the council's dual interests in underwriting marine science and permanently protecting habitat lands.

— *Rosanne Pagano*

Bibliography

Alaska Geographic Society. *Kodiak*. Vol. 19, No. 3. Anchorage, 1992.

Etkin, Dagmar Schmidt. *International Oil Spill Statistics*. Arlington, Mass.: Cutter Information Corp., 1997.

Coates, Peter A. *The Trans-Alaska Pipeline Controversy: Technology, Conservation and the Frontier*. Bethlehem, Penn.: Lehigh University Press, 1991.

Davidson, Art. *In the Wake of the Exxon Valdez*. San Francisco: Sierra Club Books, 1990.

Fried, Neal and Windisch-Cole, Brigitta. "Prince William Sound 10 years After the Spill: An Economic Profile," Juneau: *Alaska Economic Trends*, March 1999.

Exxon Valdez Oil Spill Trustees. *State-Federal Natural Resource Damage Assessment Plan for the Exxon Valdez Oil Spill*. Juneau: 1989.
—. *Exxon Valdez Oil Spill Symposium, Abstract Book*. Anchorage: 1993.
—. *Status Report*. Anchorage: 1994 through 1999.
—. *Exxon Valdez Oil Spill Restoration Plan*. Anchorage: 1994.
—. *Restoration Update*. Anchorage: March-April, 1998.

Keeble, John. *Out of the Channel: The Exxon Valdez Oil Spill,* New York, New York: HarperCollins Publishers. 1991.

Lebedoff, David. *Cleaning Up: The Story Behind the Biggest Legal Bonanza of Our Time*. New York, New York: The Free Press, 1998.

Lord, Nancy. *Darkened Waters*. Homer: Pratt Museum, 1992.

Markle, Sandra. *After the Spill: The Exxon Valdez Disaster, Then and Now*. New York, New York: Walker Publishing Co., 1999.

Mickelson, Belle et al. *Alaska Oil Spill Curriculum*. Cordova, Alaska: Prince William Sound Science Center, Prince William Sound Community College, 1990.

Mitchell, John G. "In the Wake of the Spill." Washington, D.C.: *National Geographic*, March 1999.

Monroe, Bill. "The Healing of Time, the Lingering of Wounds." Oregon: *The Oregonian*, Feb. 21, 1999.

Nalder, Eric. *Tankers Full of Trouble*. New York, New York: Grove Press, 1994.

National Transportation Safety Board. *Marine Accident Report: Grounding of the U.S. Tankship Exxon Valdez*. Washington, D.C., 1990.

National Research Council. *Tanker Spills: Prevention by Design*. Washington, D.C.: National Academy Press, 1991.

Nussbaum, Paul. "The Sound and the Fury." *Inquirer Magazine*, Philadelphia Newspapers Inc., March 7, 1999.

Piper, Ernest. *The Exxon Valdez Oil Spill, Final Report, State of Alaska Response*. Anchorage: Alaska Dept. of Environmental Conservation, 1993.

Rice, Stanley D., editor, et al. *Proceedings of the Exxon Valdez Oil Spill Symposium*. American Fisheries Society, 1993.

Romano-Lax, Andromeda. "Paradise Regained? An Oil Spill Retrospective." Anchorage: *Alaska* magazine, March 1999.

Seitz, Jody. *Alaska Coastal Currents*. Anchorage: Exxon Valdez Oil Spill Trustees, 1997 and 1998.

Strohmeyer, John. *Extreme Conditions: Big Oil and the Transformation of Alaska*. New York: Simon & Schuster, 1993

University of Alaska Fairbanks. *Alaska in Maps: A Thematic Atlas*. Fairbanks: University of Alaska Fairbanks, Alaska Department of Education and the Alaska Geographic Alliance, 1998.

Williams, T.M. and Davis, R.W., eds: *Emergency Care and Rehabilitation of Oiled Sea Otters: A Guide for Oil Spills Involving Fur-bearing Marine Mammals*. Fairbanks: University of Alaska Press, 1995.

Wills, Jonathan and Warner, Karen. *Innocent Passage: The Wreck of the Tanker Braer*. Edinburgh: Mainstream, 1993.

Following is oil spill information on the World Wide Web, with host agencies noted. Many sites also feature useful links not listed here.

http://209.17.147.1/project/pwsound.htm (University of British Columbia Fisheries Centre)

www.adn.com and www.adn.com/evos/ (*Anchorage Daily News*)

www.alaska.net/~pwsrcac/ (Prince William Sound Regional Citizens' Advisory Council)

www.alaskasealife.org (The Alaska SeaLife Center)

www/api.org/oil spill (American Petroleum Institute)

www.arlis.org (Alaska Resources Library and Information Services)

www.alyeska-pipe.com (Alyeska Pipeline Service. Co.)

www.exxon.com and www.valdezscience.com (Exxon Corp.)

www.fakr.noaa.gov/oil (National Oceanic and Atmospheric Administration)

www.oilspill.state.ak.us (Exxon Valdez Oil Spill Trustee Council)

www.pwssc.gen.ak.us (Prince William Sound Science Center, Cordova)

Index

Afognak Island 20, 48, 49, 59
Afognak Island State Park 46, 106
Agosti, Jon 69
Alaska Department of Environmental Conservation 104, 105
Alaska Legislature 104, 105
Alaska Maritime National Wildlife Refuge 56
Alaska Native Claims Settlement Act 48, 63
Alaska Oil Spill Commission 104
Alaska oil 63, 83
  Permanent Fund 64
Alaska Predator Ecosystem Experiment 15, 107
Alaska SeaLife Center 7, 47, 62, 64, 69, 71, 79, 76-80, 108
  marine life research 79, 80
Alleva, Lisa 66
Allison Creek 36
Alutiiq Museum 57, 74, 75, 107
Alyeska Pipeline Service Co. 55, 83, 87, 91, 104, 108
Archaeological resources 30, 53, 56, 57, 64-66, 74, 75

Bald eagles 13, 30, 55, 107
  spill-related research 22, 23
Barren Islands 30, 42
Bay of Isles "Death Marsh" 20
Big Waterfall Bay 54
Billys Hole 41
Black oystercatchers 30, 31
Bligh Reef 53, 102, 103
Blundell, Gail 105
"Boom" see Oil Spills, containment boom

Capps, Kris 11
Chaney, Greg 50
Chenega Bay (village) 18, 46, 67-69, 70, 104, 106, 108
Chugach National Forest 46, 53, 54, 78
Clam Gulch 2, 59
Clams 2, 30
Common murres 30, 42, 59
Cook Inlet 86
Copepods 38
Cordova 53, 81
  post-spill economy 72
Cormorants 30, 80
  pelagic 2
  red-faced 14
Cowper, Gov. Steve 86, 102, 103
Cutter Information Corp. 70

Cutthroat trout 30
Degernes, Chris 56
Dolly Varden 30

El Nino ocean warming 30
Eleanor Island 17
Eshamy Bay 89
Esler, Daniel 11, 12, 24
Evanoff, Gail 70
Exxon Corp.
  food donations 69
  oil spill science 12
  on spill zone's recovery 17
  voluntary claims program 103
Exxon Valdez accident 83, 102
  fishery closings 7, 103
  Incident Command System 96
  impact on Alaska Natives 18, 21, 45-59, 106
  impact on subsistence food consumers 104
  litigation 104, 105
  litigation settlement 14, 46, 81, 105
  mapping guides cleanup 50
  migrating birds 7
  settlement fund and land purchases 45-59
  Shoreline Cleanup Advisory Teams 50

Fucus 30, 104

Exxon Valdez Oil Spill Trustee Council 7, 12, 45-59, 106,
  Alaska SeaLife Center 78
  Alutiiq Museum 74, 75
  community project spending 47, 56
  conservation land purchases 45-59
  General Accounting Office audit of 51
  matrix to evaluate conservation lands 47
  Qutekcak hatchery 67, 69, 107
  reserve fund 108
  waste management project 71
  Youth Area Watch 73, 107
Exxon Valdez oil
  fate of spilled oil 11, 12, 60, 61
  map 52
  oil-absorbing hair filter 66
  residual oil 17, 19
  toxicity studies on 12
Exxon Valdez tanker 86, 87, 93, 102, 103
  effort to return to Alaska service 108
  renamed SeaRiver Mediterranean 105

Green Island 86
Grillo, Mikel 2, 59
Gulf of Alaska 102
  currents map 9

Habitat
  conservation land appraisals 48-51, 57
  conservation lands purchased 45-59
  studies of 16
Harbor seals 30
  spill-related research 32, 33, 79
Harlequin ducks 11, 12, 23, 30
  cytochrome P4501A 24
  spill-related research 23, 24
Harriman Fiord 22, 44, 81
Hazelwood, Joseph 84, 86, 104, 105, 108
Heintz, Ron 12, 17
Herring, Pacific 30, 31, 42, 76, 106
  lesions 37
  spill-related research 37-41
  virus 41
Homer 55

Icy Bay 97
Injured species and resources 5, 19, 22-43, 46
  at-a-glance list 30, 31
International Oil Spill Conference 101

Intertidal communities 16, 30

Jackpot Bay 45, 53
Johnson, Dr. Virginia 69

Kachemak Bay State Park 106
Karluk River 57
Kenai Fjords National Park 46, 56, 108
Kenai Mountains 97
Kenai National Wildlife Refuge 56
Kenai Peninsula 11, 55, 56, 71, 103
Kenai River 38, 53
  spill-related restoration 55, 56
Killer whales 12, 31, 108
  dorsal fins 24, 25
  resident pods 24
  social structure 26
  spill-related research, resident and transient groups 24-26
  transient pods 25
Kittlitz's murrelets 31, 107
Knight Island 5, 21, 103, 105, 107
Knowles, Gov. Tony 2, 89
Kodiak Island
  archipelago 56-59, 71, 74, 75
Kodiak National Wildlife Refuge 56, 59, 107

preservation of Karluk and Sturgeon rivers 57

Laktonen, LaRita 74
LaTouche Island 103
Littleneck clams 67, 69, 70
Loons 31, 107
Lord, Nancy 69

Main Bay 95
Malone, Bob 87, 89, 92, 95
Marbled murrelets 17, 31
Matkin, Craig 25
McCammon, Molly 12, 16, 47, 54
McCrory, Phillip 66
McDowell, Sandee 66, 96
Mearns, Allen 101
Michel, Jacqui 101
"Mousse" 98, 102
Mulcahy, Dan 11, 12
Murkowski, U.S. Sen. Frank 45, 51
Murphy, Dr. Joyce 69
Mussels 31
    cleansing of 34, 35
    spill-related research 33-35

National Research Council 83, 87
National Transportation Safety Board 83, 86, 103
Nearshore Vertebrate Predator Project 15, 29, 108
Northwest Bay 2
Nuka Bay 26, 107

Ocean currents 39
    map 9
Oil Pollution Act of 1990 83-101, 105, 108
    double bottoms, double hulls 91, 96

shippers' liability 86
Oil shipping 86-101
    Liberia 86
    oil tanker schematic 88
    citizens' oversight 89, 90, 105
    Ship Escort/Response Vessel System (SERVS) 92, 95
    Sullom Voe port 87
    Vessel Traffic System 87
Oil Spill Recovery Institute 16
Oil spills 70, 93, 94
    Alaska oil spills list 94
    bioremediation 104
    chemical dispersant 95, 102
    cleanup 84, 86, 91, 92
    containment "boom" 84, 95
    high pressure, hot water hosing 101, 103
    natural cleansing 101
    oil-on-water burning 98
    worldwide oil spills list 93
Oil spills, prevention 83-101
    Det Norske Veritas 91
    Practice drills 96
    Shoreline assessment 101, 106
    Tanker escorts 91
    Then and now in the Sound 85
    "Tractor" tugs 91
Oil tankers 70, 84, 88, 92
Okey, Thomas 18
Old Harbor 57
O'Meara, Mike 63, 64
Owens, Ed 101

Pacific herring see herring, Pacific
Page, David S. 21
Phillips, Natalie 45
Picou, Steven 81

Pigeon guillemots 17, 31, 79
Pink salmon 12, 30, 31, 42
    egg mortality studies 35
    life cycle of 36
    otolith marking 106, 108
    polycyclic aromatic hydrocarbons, sensitivity to 36
    spill-related research 35, 36, 79
Piper, Ernest 83, 96, 99
Plankton (also see zooplankton) 106
    abundance of 43
    role in food chain 42
    spill-related research 41-43
Polycyclic aromatic hydrocarbons, concentration in water 36
Port Dick Creek 53
Pratt Museum 63, 103
    "Darkened Waters" display 64, 69
Prince William Sound 8, 15, 16, 18, 19, 47, 92
    archaeology 65, 66
    commercial fishing 51, 108
    conservation lands purchased in 51-56, 107
    food chain 13, 17
    plankton's role 41-43
    practice spill drill 83
    recovery of 17
    spill-related oceanography of 39, 42
Prince William Sound Regional Citizens' Advisory Council 19, 89, 90
Prince William Sound Science Center 16, 39

Quayle, Vice President Dan 103
Qutekcak hatchery 67, 69, 70, 107

Reger, Douglas 65
Resource Development Council 46
River otters 13, 31, 105
    declared a recovered species 108
    spill-related research 26, 27, 79
Robbins, U.S. Coast Guard Vice Adm. Clyde 103
Rockfish 31, 77
Rosenberg, Dan 17, 19, 20, 24

Sale, David 99
Sampson, Roger 71
Sawmill Bay 7, 102
Sea otters, 13, 31, 107
    cleaning of 76
    spill-related research 27-29
    treatment and release following spill 29
Sea stars 11, 16, 71
Sediments 31
Seward 62, 78
Shuyak Island State Park 46, 107
Simpson, Sherry 5, 7
Sitkalidak Island 57
Sleepy Bay 103
Smith Island 103
Smithsonian Institution 64, 74, 75, 104
Sockeye salmon 31, 53, 107
Sound Ecosystem Assessment Project 15, 106
Spies, Robert 21
Steffian, Amy 74, 75
Steller sea lions 43, 69
    fisheries and 79, 80

Stephens, Stan 19
Sturgeon River 57
Subsistence foods 66
Subtidal communities 16, 31
Sundberg, Kim 76

Tatitlek 69, 90, 104
Totemoff, Chuck 54
Tourism 18, 19, 64, 65, 99
Trans-Alaska pipeline 63, 83
Troll, Ray 87

U.S. Coast Guard 84, 92, 96
Unakwik Inlet 32

Valdez 2, 78, 83, 100, 102
    bird rescue 24
    sea otter rescue 27
Valdez Narrows 87, 92, 95

Whittier 7, 18
Wiens, John A. 13
Wilderness 30
Williams, Deborah 46, 59
Wohlforth, Charles P. 60, 61

Yost, U.S. Coast Guard Commandant Paul 104
Youth Area Watch 70, 71-76, 107,

Zencey, Matt 45
Zooplankton 38, 39
    spill-related research 41-43

PHOTOGRAPHERS

The North Slope, Vol. 1, No. 1. Out of print.
One Man's Wilderness, Vol. 1, No. 2. Out of print.
Admiralty...Island in Contention, Vol. 1, No. 3. $19.95.
Fisheries of the North Pacific, Vol. 1, No. 4. Out of print.
Alaska-Yukon Wild Flowers, Vol. 2, No. 1. Out of print.
Richard Harrington's Yukon, Vol. 2, No. 2. Out of print.
Prince William Sound, Vol. 2, No. 3. Out of print.
Yakutat: The Turbulent Crescent, Vol. 2, No. 4. Out of print.
Glacier Bay: Old Ice, New Land, Vol. 3, No. 1. Out of print.
The Land: Eye of the Storm, Vol. 3, No. 2. Out of print.
Richard Harrington's Antarctic, Vol. 3, No. 3. $19.95.
The Silver Years, Vol. 3, No. 4. $19.95.
Alaska's Volcanoes, Vol. 4, No. 1. Out of print.
The Brooks Range, Vol. 4, No. 2. Out of print.
Kodiak: Island of Change, Vol. 4, No. 3. Limited.
Wilderness Proposals, Vol. 4, No. 4. Out of print.
Cook Inlet Country, Vol. 5, No. 1. Limited.
Southeast: Alaska's Panhandle, Vol. 5, No. 2. Limited.
Bristol Bay Basin, Vol. 5, No. 3. Out of print.
Alaska Whales and Whaling, Vol. 5, No. 4. $19.95.
Yukon-Kuskokwim Delta, Vol. 6, No. 1. Out of print.
Aurora Borealis, Vol. 6, No. 2. $19.95.
Alaska's Native People, Vol. 6, No. 3. Limited.
The Stikine River, Vol. 6, No. 4. $19.95.
Alaska's Great Interior, Vol. 7, No. 1. $19.95.
Photographic Geography of Alaska, Vol. 7, No. 2. Limited.
The Aleutians, Vol. 7, No. 3. Out of print.
Klondike Lost, Vol. 7, No. 4. Out of print.
Wrangell-Saint Elias, Vol. 8, No. 1. $21.95.
Alaska Mammals, Vol. 8, No. 2. Out of print.

The Kotzebue Basin, Vol. 8, No. 3. Out of print.
Alaska National Interest Lands, Vol. 8, No. 4. $19.95.
Alaska's Glaciers, Vol. 9, No. 1. Revised 1993. $19.95.
Sitka and Its Ocean/Island World, Vol. 9, No. 2. Limited.
Islands of the Seals: The Pribilofs, Vol. 9, No. 3. $19.95.
Alaska's Oil/Gas & Minerals Industry, Vol. 9, No. 4. $19.95.
Adventure Roads North, Vol. 10, No. 1. $19.95.
Anchorage and the Cook Inlet Basin, Vol. 10, No. 2. $19.95.
Alaska's Salmon Fisheries, Vol. 10, No. 3. $19.95.
Up the Koyukuk, Vol. 10, No. 4. $19.95.
Nome: City of the Golden Beaches, Vol. 11, No. 1. $19.95.
Alaska's Farms and Gardens, Vol. 11, No. 2. $19.95.
Chilkat River Valley, Vol. 11, No. 3. $19.95.
Alaska Steam, Vol. 11, No. 4. $19.95.
Northwest Territories, Vol. 12, No. 1. $19.95.
Alaska's Forest Resources, Vol. 12, No. 2. $19.95.
Alaska Native Arts and Crafts, Vol. 12, No. 3. $24.95.
Our Arctic Year, Vol. 12, No. 4. $19.95.
Where Mountains Meet the Sea, Vol. 13, No. 1. $19.95.
Backcountry Alaska, Vol. 13, No. 2. $19.95.
British Columbia's Coast, Vol. 13, No. 3. $19.95.
Lake Clark/Lake Iliamna, Vol. 13, No. 4. Limited.
Dogs of the North, Vol. 14, No. 1. $21.95.
South/Southeast Alaska, Vol. 14, No. 2. Limited.
Alaska's Seward Peninsula, Vol. 14, No. 3. $19.95.
The Upper Yukon Basin, Vol. 14, No. 4. $19.95.
Glacier Bay: Icy Wilderness, Vol. 15, No. 1. Limited.
Dawson City, Vol. 15, No. 2. $19.95.
Denali, Vol. 15, No. 3. $19.95.
The Kuskokwim River, Vol. 15, No. 4. $19.95.
Katmai Country, Vol. 16, No. 1. $19.95.
North Slope Now, Vol. 16, No. 2. $19.95.
The Tanana Basin, Vol. 16, No. 3. $19.95.
The Copper Trail, Vol. 16, No. 4. $19.95.
The Nushagak Basin, Vol. 17, No. 1. $19.95.
Juneau, Vol. 17, No. 2. Limited.
The Middle Yukon River, Vol. 17, No. 3. $19.95.
The Lower Yukon River, Vol. 17, No. 4. $19.95.
Alaska's Weather, Vol. 18, No. 1. $19.95.
Alaska's Volcanoes, Vol. 18, No. 2. $19.95.
Admiralty Island: Fortress of Bears, Vol. 18, No. 3. $21.95.
Unalaska/Dutch Harbor, Vol. 18, No. 4. $19.95.
Skagway: A Legacy of Gold, Vol. 19, No. 1. $19.95.
Alaska: The Great Land, Vol. 19, No. 2. $19.95.
Kodiak, Vol. 19, No. 3. Limited.
Alaska's Railroads, Vol. 19, No. 4. $19.95.
Prince William Sound, Vol. 20, No. 1. $19.95.
Southeast Alaska, Vol. 20, No. 2. $19.95.
Arctic National Wildlife Refuge, Vol. 20, No. 3. $19.95.
Alaska's Bears, Vol. 20, No. 4. $19.95.
The Alaska Peninsula, Vol. 21, No. 1. $19.95.
The Kenai Peninsula, Vol. 21, No. 2. $19.95.

People of Alaska, Vol. 21, No. 3. $19.95.
Prehistoric Alaska, Vol. 21, No. 4. $19.95.
Fairbanks, Vol. 22, No. 1. $19.95.
The Aleutian Islands, Vol. 22, No. 2. $19.95.
Rich Earth: Alaska's Mineral Industry, Vol. 22, No. 3. $19.95.
World War II in Alaska, Vol. 22, No. 4. $19.95.
Anchorage, Vol. 23, No. 1. $21.95.
Native Cultures in Alaska, Vol. 23, No. 2. $19.95.
The Brooks Range, Vol. 23, No. 3. $19.95.
Moose, Caribou and Muskox, Vol. 23, No. 4. $19.95.
Alaska's Southern Panhandle, Vol. 24, No. 1. $19.95.
The Golden Gamble, Vol. 24, No. 2. $19.95.
Commercial Fishing in Alaska, Vol. 24, No. 3. $19.95.
Alaska's Magnificent Eagles, Vol. 24, No. 4. $19.95.
Steve McCutcheon's Alaska, Vol. 25, No. 1. $21.95.
Yukon Territory, Vol. 25, No. 2. $21.95.
Climbing Alaska, Vol. 25, No. 3. $21.95.
Frontier Flight, Vol. 25, No. 4. $21.95. Our 100th Issue!

PRICES AND AVAILABILITY SUBJECT TO CHANGE

Membership in The Alaska Geographic Society includes a subscription to *ALASKA GEOGRAPHIC*®, the Society's colorful, award-winning quarterly.

Call or write for current membership rates or to request a free catalog. *ALASKA GEOGRAPHIC*® back issues are also available (see above list). **NOTE:** This list was current in mid-1999. If more than a year or two has elapsed since that time, contact us before ordering to check prices and availability of back issues, particularly books marked Limited.

When ordering back issues please add $4 for the first book and $2 for each additional book ordered for Priority Mail. Inquire for non-U.S. postage rates. To order, send check or money order (U.S. funds) or VISA/MasterCard information (including expiration date and your phone number) with list of titles desired to:

ALASKA GEOGRAPHIC.

P.O. Box 93370 • Anchorage, AK 99509-3370
Phone: (907) 562-0164 • Fax (907) 562-0479